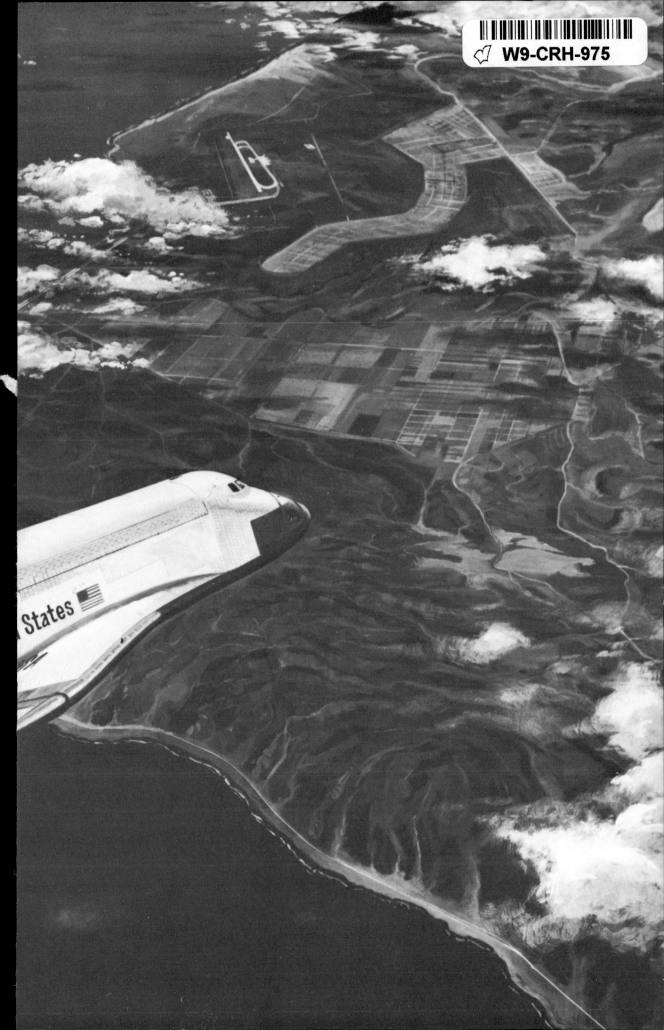

SPACE SHUTTLE

Revised Second Edition 1983

Library of Congress Cataloging in Publication Data
Kaplan, Marshall H.
  Space shuttle

  1. Reusable space vehicles. I. Title.
TL795.5.K36 629.44'9 82-70978
ISBN 0-8168-8450-1

International Standard Book Number 0-8168-8451-X
Library of Congress Card Number 82-70978

Printed and Published in the United States by Aero Publishers, Inc.

Space Shuttle

America's Wings To The Future

Second Edition

Marshall H. Kaplan

AERO PUBLISHERS, INC.

329 West Aviation Road, Fallbrook, CA 92028

To the memory of Howard S. Seifert,
my teacher and a pioneer in space technology

PREFACE

*"The key to future exploration and
use of space is the reusable earth-to-orbit
transport system"*

Wernher von Braun, 1972

A great deal has happened in the development of the Space Transportation System since the writing of the first edition of this book. What is presented here is an extensive revision and updating of that edition, and new photographs taken from actual launches have been included where appropriate. Complete chapters have been rewritten to reflect these changes. The Space Transportation System has now become operational after many years of development and testing. Soon landings will be a regular event at the Kennedy Space Center in Florida. The flight test program consisting of four launches and landings has been relatively unhindered and very successful. Soon America will send its first woman astronaut into space in one of the shuttle flights, and commercial payloads will be taken on as cargo. Space stations will be built for military and civilian uses. Bigger and bigger satellites will be launched because of the advantages gained by the Space Transportation System.

All these events have been the culmination of a great deal of work on the part of hundreds of thousands of people over the past decade. Space technology is changing, ever so subtly, the way of life in America and the rest of the world. For example, we will soon have electronic mail, direct reception of television from satellites with a very small roof antenna, and possibly high-powered laser devices in orbit for the defense of the United States.

Yes, it is time for a second edition to the Space Shuttle book. Intense rewriting of this edition took place over the early months of 1982. Each flight of the Space Shuttle was followed and incorporated into the text material. Therefore, this book represents the latest information on the Space Transportation System.

Many people have assisted in generating this new edition. I would particularly like to thank certain of the clerical and professional staff at Spacetech: Alice Bigatel, whose capabilities and talents with her typewriter made the process of revision possible; Mary D'Urso, Kay Romano, and Patricia Szybist, all of whom shared in the labors of conscientious review and correction of the final copy; and Maureen Glancy, who managed the overall production task and provided editorial services at all stages of development.

Again, I certainly hope you enjoy reading this book as much as I enjoyed writing it.

Marshall H. Kaplan
State College
Pennsylvania

INTRODUCTION

BIRTH OF A NATIONAL SPACELINE

It is 6:55 in the morning on April 12, 1981. The sun has risen and brought to light clear, blue skies with cool morning breezes to Cape Canaveral, Florida. You are there to witness a historical event which will take place within the next few minutes. This will be the launch of the first Space Shuttle into orbit. After almost nine years of intense development and testing, the National Aeronautics and Space Administration is about to try out their new system for transporting cargo into orbit, the *Space Transportation System.* We know it as the *Space Shuttle.*

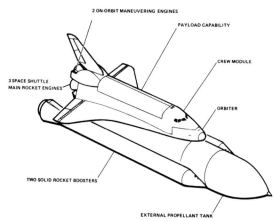

The Space Shuttle flight system consists of the Orbiter, an External Tank, and two Solid Rocket Boosters. The Orbiter contains three main rocket engines and two smaller motors for orbital maneuvering.

Let's look back for a moment and see how we got to this point. On December 17, 1903, Orville and Wilbur Wright were successful in achieving sustained flight in a powered aircraft. There were only a few spectators that cold, gray December morning in 1903. The first flight lasted only twelve seconds over a distance of 120 feet. They demonstrated an average speed of thirty-one miles per hour. The Space Shuttle *Orbiter,* named *Columbia,* which is the airplane-like craft you are about to see launched, is the length of the Wright brothers' historic first flight, but it will achieve speeds of 17,500 mph as it circles the earth. Historical events are not always instantly acclaimed or spectacular. The initial notification of the Wright brothers' success was simply a telegram to their father. Sixty-six years later, a man first stepped onto the

The Orbiter is illustrated in a cutaway manner to expose the various components and crew positions which make up this complicated combination spacecraft and aircraft. It is launched vertically as part of the Space Shuttle vehicle and lands on a runway, similar to those used by present-day jet aircraft.

moon, and an estimated 500 million people around the world saw the event on television or listened to it on the radio as it happened. Neil Armstrong proclaimed, "One small step for a man. One giant step for mankind." We are about to witness another giant step, for on this morning in 1981, the first Shuttle flight will be seen and heard around the world as it happens. You will be sharing the enthusiasm of almost a billion people as the five rocket motors scream upwards through the atmosphere and on into space.

Five . . . four . . . three . . . two . . . one . . . zero. The huge engines light up, and a mighty roar is heard. Within seconds, the big bird will leave historic pad 39A from which the Apollo moon rockets were launched. All five gigantic engines will roar at full blast. At first, the huge piggyback combination of rockets, tank, and Orbiter will move ever so slowly upward. As it clears the gantry, you will notice the vehicle turn about its axis and move perceptibly faster. Seconds later, it will be accelerating, and its flight will take the bird out toward the eastern horizon. Within a few minutes, Columbia will be hurled up to speeds of over 17,000 mph into orbit around our globe. On the way up, it will drop two huge rocket motors into the Atlantic Ocean and, later, a gigantic propellant tank into the Indian Ocean.

This was the first test flight of the system. There were a total of four such flights between Spring 1981 and mid-1982. A whole new era of transportation then came into being with the inauguration of an operational system in 1982. This is known as the *National Spaceline,* providing almost a monthly delivery service for payloads going into orbit. The system is planned to consist of up to five cargo carriers (Orbiters) with associated boosters and tanks, two launch and landing sites, supporting facilities, and an assortment of devices for on-orbit services. Both men and women crewmembers can be aboard the Shuttle. The Spaceline provides delivery of satellites, platforms, and supplies, and return of old spacecraft. What makes the National Spaceline possible is that the system is largely reusable and offers a variety of options for users. It will lead to the economical and routine use of space. Hopefully, space will now become commonplace.

The Space Shuttle has come. Just as scheduled airline operations started 22 years after the first flight in 1903, the National Spaceline started scheduled service 24 years after our historic first flight of Explorer I on January 31, 1958. Over 2,000 spacecraft have been placed into orbit by throwaway launchers. However, the Space Shuttle is basically a reusable system. It leaves Earth

In 1947 the Sacramento Bee *printed this article entitled "A Trip to the Moon and Back." This represents some early serious thoughts on space travel. Although the Space Shuttle does not use the launch technique depicted here, people are again thinking about launching future Shuttles from large aircraft. There are a number of similarities between the spacecraft shown here and the Orbiter.*

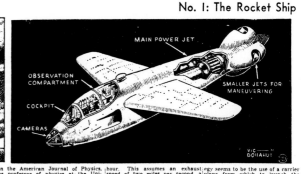

THE SACRAMENTO BEE, THURSDAY, FEBRUARY 13, 1947

A Trip To The Moon And Back

No. I: The Rocket Ship

like previous boosters (vertically from a launch pad), but the Orbiter returns like an airplane.

Let's look back into history once again. The concept of employing aircraft and booster technology is being used to bring about a versatile and economic transportation system for space. This is not a new idea, however. In 1947, a newspaper article appeared in the *Sacramento Bee,* entitled "A Trip to the Moon and Back," in which Professor Henry A. Erikson speculated about a lunar rocket launched from a huge airplane. In 1954, *Colliers Magazine* carried articles written by a committee of experts, including Wernher von Braun, concerning the possibility of a reusable earth-to-orbit space transportation system. Technology has advanced by leaps and bounds over the last twenty years. It is now time for the Space Shuttle. Indeed, this is an idea whose time has come.

Completion of the Apollo Program served as a cue to the next logical step in making space-flight economical and accessible to a wide variety of users. In the '60s and '70s it was men going to the moon to prove that it could be done. In the '80s and '90s, it will be men and women going into space for the good of all mankind. On August 9, 1972, the National Aeronautics and Space Administration authorized Rockwell International's Space Division to proceed with the Space Shuttle Orbiter. Several other system components were contracted directly to NASA at about that time. These included construction of launch facilities at the Kennedy Space Center, Florida, production of propellant tanks, and development of rocket boosters and shuttle main engines. All these activities climaxed in the first orbital flight this historic morning in April 1981.

The primary goal of the Space Shuttle Program is to provide low-cost transportation to and from earth orbit. Scientific laboratories will be carried aloft in support of staffed experiments in space. Free-flying

The Space Transportation System has truly been a national project. Over 200 companies in 44 states have received contracts related to different parts of the system. This map represents a graphic illustration of the distribution of spending. The superimposed numbers tell us how many millions of dollars were planned to be spent in each of these states.

satellites will be deployed and recovered. Robot-like satellites with rocket stages attached will be released from the Shuttle and sent into high-energy trajectories, many of them escaping the pull of earth's gravity altogether to journey to other planets and asteroids.

The Space Shuttle has been built by many industrial contractors around the country who are responsible for the major components. For example, Rockwell International received large contracts for the Orbiter and its main rocket engines. At least twenty other contractors received orders in excess of ten million dollars each. Some 200 companies in 44 states have received contracts to provide hardware and services for the new Space Transportation System. The distribution of spending in this program has truly been on a nationwide basis.

What you are about to read is the how, what, when, and why of the Space Shuttle. You will learn all about the different parts of the system and the schedule which brought about the inauguration of the National Spaceline. Every phase of the mission will be described, including preparation and training of the crew, loading of the payloads, the sensation of liftoff and acceleration upward, and extravehicular activities in orbit. Activities begin about two weeks before liftoff and proceed through the orbital flight and return to earth. Benefits to you, as to all peoples, are reviewed, both from the introduction of a National Spaceline and from spaceflight in general. Yes, the Space Shuttle is for you and yours.

A partial list of companies receiving contracts for Space Shuttle work appears in Appendix A. Management of the program is shared by NASA headquarters and three NASA centers. The system elements were supplied to NASA by prime contractors, assisted by many subcontractors.

ORBITER
PRIME CONTRACTOR—Rockwell International, Space Division, Downey, Calif.
SUBCONTRACTS
Wing—Grumman Aerospace, Bethpage, N. Y.
Vertical Tail—Fairchild Republic, Farmingdale, N. Y.
Mid Fuselage—General Dynamics, San Diego, Calif.
Reusable Surface Insulation—Lockheed Missiles and Space Co., Sunnyvale, Calif.
Orbital Maneuvering System—McDonnell Douglas Astronautics Co., E. St. Louis, Mo.
Orbital Maneuvering Engines—Aerojet Liquid Rocket Co., Sacramento, Calif.

EXTERNAL TANK
PRIME CONTRACTOR—Martin Marietta Aerospace, Denver, Colo.
SUBCONTRACTS
Intertank—Avco Corp., Nashville, Tenn.
Dome Caps—General Dynamics, San Diego, Calif.
Ogive and Dome Gores—Aircraft Hydroforming, Gardena, Calif.
Tank Weld Tools—LTV Aerospace, Dallas, Tex.
Gore/Dome Weld Tools—The Boeing Company, Seattle, Wash.

SOLID ROCKET BOOSTER
PRIMARY MANAGEMENT—Marshall Space Flight Center, Huntsville, Ala.
SUBCONTRACTS
Assembly and Checkout—United States Boosters Inc., Huntsville, Ala.
Solid Rocket Motor—Thiokol Corp., Brigham City, Utah
Structures—McDonnell Douglas Astronautics Co., Huntington Beach, Calif.
Decelerator Subsystem—Martin Marietta Aerospace, Denver, Colo.
Integrated Electronics Assembly—Bendix Corp., Teterboro, N. J.
Booster Separation Motor—United Technology Corp., Sunnyvale, Calif.

SPACE SHUTTLE MAIN ENGINE
PRIME CONTRACTOR—Rockwell International, Rocketdyne Div., Canoga Park, Calif.
SUBCONTRACTS
Controller—Honeywell Inc., Minneapolis, Minn.
Hydraulic Actuator System—Hydraulic Research, Valencia, Calif.

FOREWORD

For centuries man had looked longingly at the heavenly scene and dreamed of being able to go to those distant planets. He created myths about Daedalus and Icarus, who flew on gossamer wings. Through the talents of Leonardo da Vinci, he designed imaginary airplanes and space ships. Through many outstanding writers he has vicariously traveled through space with Buck Rogers, Flash Gordon, and a host of other "spacemen."

Then in 1957 the orbiting of Sputnik brought the dreams to the brink of possibilities. These dreams began to materialize into realities with the earth orbiting of Soviet cosmonauts and American astronauts. Consequentially, the dreams and challenges of space, as well as international competition, spurred man on to make the first explorations of another heavenly body when our American astronauts walked on the surface of the moon.

Suddenly man had developed the space craft, the propulsion system, the communications system, and the navigation system that could take him not only to the moon, but to Mars, or Venus, or Jupiter, or even beyond. Thus, entire new vistas were opened that were no longer distant dreams but now very real possibilities.

Of course, there were limitations to these vistas—and that is the problem of cost. Man now found himself debating, "Is space worth the cost?" The answer came back that the direct and indirect benefits from the rapid achievement of technology were already showing high yields on dollars invested—but could man continue to invest such sizable amounts? The Soviets felt yes, the Americans felt no, at least not right now.

So now we ask, when will the national mood be receptive toward space exploration and industrialization? Will the United States wait until Russia spurs her on by more outstanding accomplishments? Or, will she start a logical and orderly program of exploration? It seems unbelievable that the American public has approved many billions of dollars of giveaway (welfare) programs that have had minimal dollar return, but have been indifferent toward **high technology space programs which have already yielded over 24 dollars for each and every dollar invested.** Furthermore, these space programs have accelerated all fields of technology and have gainfully employed hundreds of thousands of American citizens.

It is beginning to appear that the public's mood will be receptive to various satellite services that provide direct benefits to large numbers of people, and also help solve great national and global problems. They will probably be even demanding of those services that provide for personal safety, personal communications, personal navigation, improvement in medical care and improvement in environment.

Technology will not be the limiter of such advances, but return on investments will be. Therefore, the industrial exploration of space can be accomplished by the Space Shuttle. The cost per pound of payload will come down with weekly launches, and the monetary yields will increase to the point where industry will be competing to get into the various Space Shuttle projects. Already other countries and their industries are eagerly joining in participating in what will very likely become a true world space project. Hopefully, the enthusiasm and support of the American public will increase to the point of backing this project as we move toward lunar and asteroid mining, large scale manufacturing in space, space colonization and scientific manned exploration of our own solar system as well as neighboring systems.

Gordon Cooper

TABLE OF CONTENTS

CHAPTER ONE

PREPARATIONS FOR A NEW TRANSPORTATION SYSTEM

WHAT IT TOOK TO GET THE SHUTTLE STARTED

In mid-1972, the National Aeronautics and Space Administration announced the selection of Rockwell International to be the builder of the Orbiter. The envied award of the primary contract for the new Space Transportation System was preceded by years of study, preliminary design, and proposal writing. Four giant aerospace companies were vying for this important contract: Grumman, Lockheed, Rockwell, and McDonnell Douglas. Each had spent millions of dollars in developing their own designs and generating a proposal of monumental proportions. Imagine the scene as a huge delivery truck pulls up to NASA Headquarters in Washington, D.C. to deliver the four proposals in 1971. Each proposal contained thousands of pages of design, management, schedule, and cost information. Each one filled a large filing cabinet. Imagine having to read through these expansive documents and come up with an evaluation within just a few months. This was NASA's job in 1971. Remember, this proposal was for the Orbiter only. The contract did not include other large components such as the main rocket engines, boosters, and external propellant tanks.

Of course, when you realize that the Shuttle will replace all other U.S. launch vehicles, this tremendous effort was justified. After all, the Shuttle cost the taxpayers several billion dollars before its first flight, and we want to be sure to get our money's worth. Judging from our earlier investments in space, we will indeed reap benefits far in excess of our initial investment. My confidence is reaffirmed every year when Congress appropriates the annual budget for the Shuttle. Its investigative arm, the U.S. General Accounting Office watches over the program to insure the most prudent expenditure of our tax dollars so we are assured that budgetary objectives as well as technical ones are being satisfied by NASA.

Over the years since 1958 there have been periodic outcries that our spending in space is wasted. This was especially true during the Apollo moon flights when people accused NASA of spending billions on the moon while thousands of people in this country were starving. Of course, this is a ridiculous accusation. Every penny ever spent on space is right here on earth where hundreds of thousands of jobs in everything from high technology to janitorial services have been created. This will also hold true for the Shuttle and, in fact, more so. One of the primary intentions for developing the Shuttle is to make space more usable for commercial ventures.

New Year's Eve 1971 was approaching quickly when the then NASA Administrator, Dr. James C. Fletcher, was still negotiating with the President's Office of Management

Space Shuttle Orbiters are assembled at the U.S. Air Force Plant #42 at Palmdale, California. This facility is used by Rockwell's Space Division for the final assembly, test, and checkout of the vehicles. This picture was taken during the rollout ceremonies of the Enterprise on September 17, 1976.

and Budget. They were in the process of thrashing out the initial level of spending needed to get the Shuttle started in 1972. Alas, it is the president who must request funds from Congress for such programs. At one point in these negotiations, budget officials began suggesting design changes in the system which drew Fletcher's fire, claiming that the Office of Management and Budget was attempting to design the U.S. Space Shuttle rather than merely to budget it. President Nixon agreed with Fletcher and directed the budget office to settle the arguments quickly. By the first week in January 1972, the budget was settled and the Shuttle funds became part of the president's request to Congress. Later that year, the U.S. Congress approved these expenditures, and Rockwell was given the go-ahead for the Orbiter construction on August 9, 1972.

A Harris poll taken in 1981 indicated how Americans felt about spending several billion dollars during the 1980s on the Space Shuttle Program. The results show strong support, even though the nation was in a budget cutting mood at that time. The question asked of 1,250 people was: "It could cost the U.S. government several billion dollars to develop the full potential of the shuttle over the next 10 years. All in all do you feel this space program is worth it?"

	Yes	No	Not Sure
Men	76%	21%	3%
College educated	71%	26%	3%
Women	52%	43%	5%
Blacks	45%	53%	2%
Total	63%	33%	4%

NINE YEARS OF HARD WORK

About the same time that Rockwell began building the Shuttle Orbiter, NASA was negotiating and initiating several other major contracts for different parts of the Space Transportation System, including main rocket engines, boosters, and propellant tanks. All elements of the system had to be developed on a schedule that would bring everything together for that first launch on April 12, 1981. It took almost nine years of hard work before the Shuttle reached into space with that first orbital test launch.

All elements except the main engines moved toward completion smoothly. The main engine development was an extremely critical part of the overall system schedule. The contract for these engines alone totalled over one billion dollars. However, troubles with the main rockets developed during test firings. Pumps needed to supply liquid oxygen and liquid hydrogen to the thrust chambers at higher-than-previous pressures burned out during various critical test firings. This contributed to significant cost overruns and delays in many program areas. Back in 1971 the total cost of developing the Shuttle program over the 1972 to 1980 period was estimated to be 5.2 billion dollars. Cost overruns inflated this figure to 10 billion dollars due to technical problems, funding delays, and additional requirements for backup equipment.

In a tremendously complex program such as the Space Shuttle, there were literally thousands of hurdles to be overcome. Prior to the first flight, the approach and landing tests were the most exciting and spectacular phase of the development program. Imagine the sensation of being in the Shuttle Orbiter the first time it separated from the 747 over the dry lake bed in California. This was the climax to a 45 minute flight in which the craft was mounted piggyback on a Boeing 747. Suddenly you are free, free to control the glider, free to steer your own course. You are at the controls of a half-billion dollar combination spacecraft and aircraft. It is like nothing that has ever flown before. Your sink rate is 12,500 feet per minute, about 20 times normal. In fact, the whole flight from 20,000 feet to the desert takes only about two minutes. It is true that, if you could throw a dead body out at 10,000 feet, the Orbiter would reach the ground first.

The vehicle used in these approach and landing tests is known as Orbiter Vehicle 101 or *Enterprise*, whose rollout took place on September 17, 1976, in Palmdale, California. In order to mate the Orbiter with the 747, the spacecraft had to be moved to the Dryden Flight Research Center, about 36 miles from Palmdale. Early in the morning of January 31, 1977, Orbiter 101 was eased out onto the highway between Palmdale and the Dryden Flight Research Center. It took about twelve hours for this short trip. The Orbiter was mounted on a trailer which had 90 wheels to

The Rocketdyne Division of Rockwell International is located in Canoga Park, California. This division was responsible for the design, development, and building of the Space Shuttle Main Engines.

insure that its ride was absolutely smooth. Several months of preparation were required to make this move because of the wide load to be carried on a truckbed. Even telephone poles and street signs were removed. At one point the Orbiter wing cleared a pole by only inches. There was a carnival-like atmosphere. Local residents and visitors followed the procession on foot and by bike and car. Refreshments were sold at various points along the way. One uninformed resident living along the route woke up that morning, looked out the window, and gasped, "Oh, no, another bad landing!"

Each Orbiter, as it is completed at Palmdale, must make this same ground trip to Dryden. After arriving at this flight research center, a huge device was used to lift the Orbiter and to mate it atop NASA's Boeing 747-100 Shuttle Carrier Aircraft in a piggyback fashion. This particular 747 was purchased from American Airlines on June 17, 1974, after being used in its fleet as a passenger airliner. NASA was able to save about half the price of

a new plane by doing this. If you look closely at the 747's fuselage, just above the red, white and blue stripes, and just in front of the wing, you may still see the faint remains of the *American* logo. After the purchase, this plane was flown to the Boeing plant where it was modified for its unique and historic mission to ferry Orbiters to and from various facilities and to release the *Enterprise* for the approach and landing tests.

After the mating of the 747 and Orbiter 101, taxi tests were carried out to determine structural loads and responses. With the completion of these tests, preparations began for five flight tests with no crew aboard the Orbiter. These flights consisted of takeoff, cruise flight, and landing with the Orbiter remaining attached to the aircraft. A special tail cone was attached to the aft section of the Orbiter to reduce drag during the flight. Later this tail cone was removed and simulated rocket engines were placed in their normal positions. Before doing this, however, many tests were conducted, including landing tests, with the

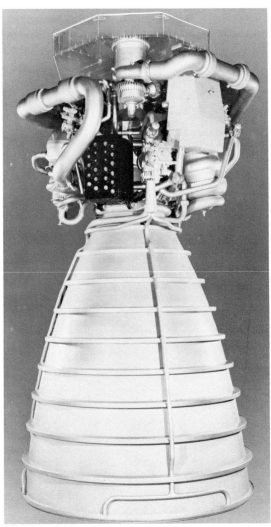

The main propulsion system includes three Space Shuttle Main Engines. These are reusable, high performance, liquid propellant rocket engines with variable thrust. They must burn continuously for eight minutes during each launch and must survive 55 starts. Each one produces a liftoff thrust of 375,000 pounds, but has a dry weight of 6,500 pounds. Each engine is 14 feet long and 8 feet in diameter.

tail cone on. Finally, on June 18, 1977, the first of three flight tests were carried out in which a two-man crew was onboard the Orbiter. The two craft remained mated during takeoff, climb, cruise, descent, and landing. During these three flights, all the various systems were checked out and evaluated in order to prepare for the upcoming separation and landing tests. To give you an idea of how complicated the system is there are five computers on each Orbiter. Four are utilized for the Orbiter primary flight control, and the fifth is for backup purposes only.

If any one of these computers fails, the mission can go on as planned. Even if a second one fails, the Orbiter will be able to return safely to earth. These innovations are typical of the safety precautions taken in building such a system. You might say that it is far safer than driving your own car down the highway.

At 8:00 a.m., Pacific Daylight Time, on August 12, 1977, the free flight phase began with a takeoff and climb to an altitude of 28,014 feet with the Orbiter tail cone on. Cruising at this altitude for approximately two minutes, the mated 747 and Orbiter 101 then pitched over into a minus seven degree descent, attaining the desired separation speed and flight path angle. The 747 reduced its engine thrust to idle and deflected its spoilers to increase drag and decrease lift. The aircraft then stabilized for separation which occurred at an altitude of 24,100 feet. At that instant the speed was 322 miles per hour. Fred W. Haise, Jr. and C. Gordon Fullerton were at the Orbiter controls. Within four seconds of separation they were 140 feet above the 747. In fact, when Astronaut Haise pushed the separation button, the Orbiter virtually "popped" smoothly off the back of the aircraft. At three seconds after release the 747 initiated a roll to the left, and the Orbiter turned to the right to provide a safe separation distance as quickly as possible. The Orbiter then pitched down, accelerated, and performed a practice flare, allowing the airspeed to decrease while evaluating flying qualities. Following this test Haise pitched the vehicle down to accelerate and at the same time initiated the first of two 90-degree turns to the left which aligned it with Runway 17 at Edwards Air Force Base. Yes the Orbiter was in fact in a standard left-hand approach pattern to the airport, even though it was still between 15,000 and 20,000 feet above the ground. By the time the second turn was completed, the Orbiter had descended to an altitude of about 6,400 feet above the ground and was about nine miles from the touchdown point.

Once lined up with Runway 17 the first approach flare maneuver started at an altitude of approximately 900 feet. This transferred the Orbiter from a glide slope of nine degrees to one of three degrees. Landing gear was deployed at about 180 feet, and the landing flare was initiated at slightly less than 100 feet altitude. The final flare establishes a sink rate of less than 10 feet per second which must be

Orbiter vehicle 101, Enterprise, *was the first vehicle off the assembly line, but it will probably never fly in space. It was used to demonstrate the approach and landing qualities of this configuration and later went through a series of structural tests. The* Enterprise *is being used for checking Orbiter compatibility at the Vandenberg facilities which are presently under development. Current NASA plans are to upgrade this vehicle for space only if needed.*

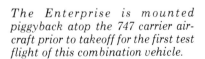

Special precautions were taken for the transport of Orbiter 101 through the community of Lancaster, California on January 31, 1977. Telephone poles, street lights, overhead wires, and road signs were moved as the 78-foot-wide, 57-foot-high spacecraft was moved from the Rockwell assembly plant to the Dryden Flight Research Center at Edwards, California. Each Orbiter, upon completion, is transported to Dryden in this manner.

The Enterprise *is mounted piggyback atop the 747 carrier aircraft prior to takeoff for the first test flight of this combination vehicle.*

Three NASA flight test crewmen were assigned to fly the 747 carrier aircraft for the Space Shuttle approach and landing tests. From left to right they are Thomas McMurtry (copilot), Fitzhugh Fulton, Jr. (pilot), and Victor Horton (flight engineer).

maintained at touchdown. The wheels made contact with the runway at a speed of 218 miles per hour. Since the tail cone tends to reduce drag and to extend the flight time, this approach and landing test took a total time of five minutes 23 seconds.

The last two flights were flown with the same configuration that the Orbiter would have on its return from space. In other words, the tail cone was taken off and replaced with three simulated main rocket engines. These flights were basically the same as the other three, except the release altitude reached was lower due to the higher drag of the Orbiter. Thus, these flights began at an altitude of about 20,000 feet.

On October 26, 1977, the last free flight occurred with Haise and Fullerton at the controls. The one minute, 59 second free-fall ride was uneventful until the final approach began. As the Orbiter passed the runway threshold, its speed was 23 miles per hour faster than planned. The split rudder speed brake was activated, and Haise tried to force the nose down in order to touch down at a specified point. As a result, the left wing dropped sharply. Haise recovered a wings-level attitude, but Orbiter 101 (the *Enterprise*) bounced about 20 feet into the air. It bounced a second time before settling onto the runway. Since it was concluded that this rough landing was due to pilot induced oscillations, the free flight tests were completed with the October 26 landing.

Orbiter 101 was then modified for a shake test at NASA Marshall Space Flight Center in Huntsville, Alabama. In March 1978, it was transported on top of the 747 to this center for testing. With the completion of those tests, the *Enterprise* was used in a number of other ground tests. Plans do not call for it to be used in actual space flights unless needed later. Extensive modifications would be required to upgrade it for flight use.

THE FINAL YEAR OF TESTING

The many years of testing, building, and assembly led to that first orbital flight in April 1981. This initiated a final year of testing before inauguration of regular cargo service into orbit. It was determined that four flight tests would be sufficient to prove the capability of the system. The first one was successfully launched on April 12, 1981; the second on November 12, 1981; the third on

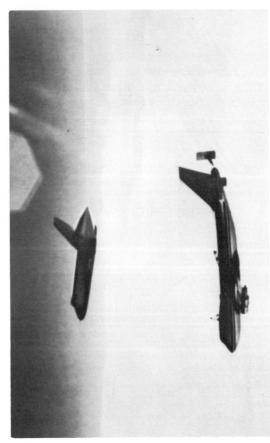

On August 12, 1977, the first free flight of the Orbiter Vehicle 101, Enterprise, took place over the Mojave Desert. Astronauts Fred Haise and Gordon Fullerton piloted the Orbiter after separation from the 747 carrier aircraft to a safe landing during the 5½-minute flight.

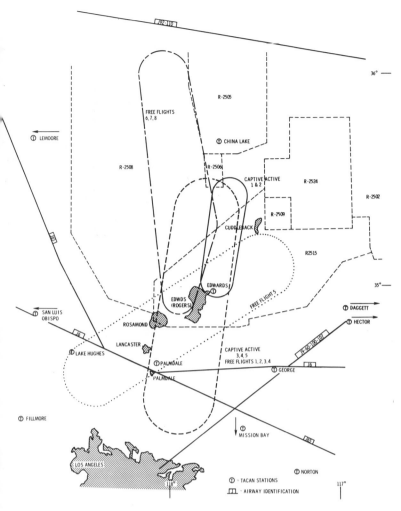

Planned ground tracks for the Shuttle Orbiter approach and landing tests are depicted in this drawing. Although the actual tests did not follow this exactly, the general flight plan is illustrated. Only five free flights were needed to verify the Orbiter's flying qualities.

March 22, 1982; and the last test flight on June 27, 1982. The first flight was totally committed to testing the system itself. Remaining test flights began carrying payloads of experiments and small spacecraft. These four journeys into space carried crews of two and landed at either Edwards Air Force Base in California or at White Sands Missile Range in New Mexico. Operational flights, i.e., those occurring after mid-1982, will land at the Kennedy Space Center in Florida. Later flights will also land at Vandenberg Air Force Base in California.

A variety of satellites and experiments will be carried up into space and some will be returned. One orbital flight test, for example, was going to carry a robot-like spacecraft to the vicinity of our old space-station, Skylab. It would have been released and commanded to dock with the massive hulk of our biggest space vehicle yet. Its purpose was to change the natural destiny of Skylab by either extending its life in orbit (raise the orbit) or cause it to reenter over a specified ocean area.

This automated spacecraft would have docked just as the previous astronauts had docked using the Apollo docking adapter. After docking, small rockets would have been fired to control the trajectory of Skylab. Unfortunately, the Shuttle was delayed and Skylab came down in July 1979.

With the completion of the last orbital flight test in 1982, the Space Shuttle system became operational, and regular service began late that same year. All of the four test flights were staffed by astronauts selected in the Apollo days and who have remained in that status. For example, John W. Young and Robert L. Crippen commanded the first flight. Young was the ninth man to walk on the moon. This was Crippen's first space flight. The second flight was commanded by Joe H. Engle and Richard H. Truly. Then Jack R. Lousma and C. Gordon Fullerton took the third flight up. The last test flight was staffed by Thomas K. Mattingly and Henry W. Hartsfield.

By the time the last flight test was completed in July 1982, two new groups of Shuttle Astronauts had been selected, trained and qualified for duty. The first group of 35 men and women were selected in December 1977, and completed a two-year training program in 1980. They were selected from over 8,000 applicants to fill positions as candidates for Pilot Astronaut and Mission Specialist Astronaut. In May 1980, nineteen more candidates were selected and trained in a condensed, one-year program. Pictures and biographical data on these candidates appear in Appendix C. Although women and men have been trained as astronauts the first female astronaut to be scheduled for space flight is Sally K. Ride. She will fulfill the role of Mission Specialist aboard the STS-7 flight in 1983.

SPACEPORTS

A major portion of the Space Transportation System is the launch and landing sites. As early as 1971 a review board was selected to search out the best candidate sites and to make recommendations concerning suitability and economic aspects of the most promising candidates. The board consisted of five dis-

Astronauts Fred Haise (left) and Gordon Fullerton walk away from the Enterprise following the third free flight of the spacecraft on September 23, 1977. The flight was uneventful and the two men said it was just another day's work.

Four astronauts were chosen to be the two crews for the Orbiter 101 Approach and Landing Tests in 1977. They are, from left to right, C. Gordon Fullerton, pilot of the first crew; Fred W. Haise, Jr., commander of the first crew; Joe H. Engle, commander of the second crew; and Richard H. Truly, pilot of the second crew.

The Shuttle carrier aircraft and six T-38 chase planes fly a formation salute over Enterprise following the third successful approach and landing test flight on September 23, 1977.

Enterprise rides piggyback on the 747 carrier aircraft for its first tail-cone-off free flight. Joe Engle and Richard Truly brought the Orbiter safely back to Earth during the two-minute-34-second free flight on October 12, 1977.

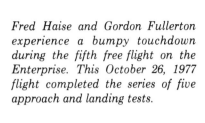

Fred Haise and Gordon Fullerton experience a bumpy touchdown during the fifth free flight on the Enterprise. This October 26, 1977 flight completed the series of five approach and landing tests.

tinguished figures associated with the Shuttle program. It was chaired by Dr. Floyd L. Thompson, Director Emeritus, Langley Research Center. The other members were Maj. Gen. Edmund F. O'Connor, Deputy Chief of Staff, Procurement and Productions, Headquarters, Air Force System Command; Mr. Vincent L. Johnson, Deputy Associate Administrator, Office of Space Science, NASA Headquarters; Mr. Robert H. Curtin, Director, Office of Facilities, NASA Headquarters; and Mr. LeRoy E. Day, Deputy Director, Space Shuttle Program, NASA Headquarters. The final site selections were to be made by the Administrator of NASA.

When the formation of this Board was publicly announced, there was an immediate influx of suggestions for these sites. Just think of the economic ramifications of having a launch site near your town. Hundreds of thousands of new jobs would be created. Money and people would flow in initially to construct this site. This inflow would level off once regular operations began. Nevertheless, there would be a definite economic advantage to having one of these sites near you. Of course there were many, many other factors to be considered in the site selection process. For example, the selected locale had to afford buffer zones surrounding the site, unpopulated areas for booster impact and safe abort maneuvers, remoteness from downrange populous areas, and a community that could accommodate all the generated activity. The booster rockets had to have at least 20 fathoms of water into which to splashdown for recovery. This was later changed to 30 fathoms or 180 feet of

water. The splashdown area was between 115 and 230 miles downrange of the launch site.

When the Board was formed the Space Shuttle System consisted of a staffed, flyable Booster and a staffed Orbiter. The Booster-Orbiter combination was to take off vertically, as a rocket, with both stages landing like airplanes. In addition, the question of whether two sites were needed was raised. This precipitated a rash of offers from all over the country. Persons representing 40 states requested that NASA locate the launch and landing site within their states. When these were added to those sites identified by NASA, a total of 150 locations had to be considered. However, on March 15, 1972, NASA changed the flyable booster concept to a rocket booster in order to minimize development costs and to stay within extreme budget constraints imposed by the Office of Management and Budget. It turned out that no inland body of water provided sufficient area for the booster impact, except possibly in the Great Lakes where community encroachment would be unacceptable. Thus, further site considerations were limited to coastal areas at that time. Screening coastal sites included consideration of booster rocket impact zones, ascent phase sonic booms, and landing field requirements. Finally, the potential sites were narrowed down to locations on the east and west coasts and the Gulf of Mexico. West Coast areas, except Vandenberg Air Force Base, were eliminated due to terrain limitations and because existing community development would impede or prevent necessary land acquisitions. East Coast sites north of

Skylab, our first space station, was launched in 1973 and was visited by three different crews. Abandoned in February 1974, this huge 85-ton hulk slowly descended back to Earth and died in a fiery reentry on July 11, 1979, after a premature rescue plan was abandoned.

Chesapeake Bay were eliminated because it was unlikely that the government could acquire sufficient land area for the site along this densely populated region. Sites in North and South Carolina apparently had clear launch directions available, but closer studies showed these mainland areas had well established communities. Thus, only Vandenberg Air Force Base and Kennedy Space Center survived as potential launch sites on the east and west coasts of the United States.

Vandenberg Air Force Base could provide for launches into orbits which pass the higher latitudes of the world. For example, *sunsynchronous* missions, which support Air Force requirements would be launched from this base. Such orbits require passage over the very high latitudes of the world. However, this site cannot provide easterly launches into orbits which only cover lower latitudes of the world. Kennedy Space Center would provide for launches with this requirement. In fact, most of the launches would take place from the Kennedy Space Center because this facilitates entry into orbits which are most popular. On the other hand, the Kennedy Space Center would not be allowed to provide

John Watts Young was born on September 24, 1930, in San Francisco. He commanded the first orbital test flight. This was his fifth ride into space. He was the pilot of Gemini 3, command pilot of Gemini 10, command module pilot of Apollo 10, and commander of Apollo 16.

Robert Laurel Crippen was born on September 11, 1937, in Beaumont, Texas. He was the pilot on the first orbital test flight. This was his first space flight.

This 1967 aerial mosaic of Merritt Island shows the Kennedy Space Center and surrounding communities. If you look closely you will see launch pads 39A and 39B in the right center of the picture.

The Space Shuttle Orbiter Enterprise rolling out of the Vehicle Assembly Building for the three and a half-mile journey to Launch Complex 39's pad A. Orbiter landing facilities are seen at top of picture in the background.

for entry into those orbits which pass over the high latitudes of the world because this would require flying over the United States on the early part of the ascent.

Upon investigating the Gulf Coast, an area was found in Matagorda County, Texas, that had the potential to accommodate much of the total program. This site apparently had cleared areas for booster impact and recovery, and the area seemed sufficiently free of existing development. Thus, the Board had narrowed down the selection to two combinations; either Kennedy Space Center and Vandenberg Air Force Base or a Gulf Coast area in Matagorda, Texas.

The Site Selection Board issued its findings on April 10, 1972. It had concluded that no existing single site could satisfy the total program requirements, and the final candidate combinations were those just mentioned. Thus, the Board gave higher ranking to the pair of currently operating launch sites, Kennedy Space Center and Vandenberg Air Force Base. Based on these findings, the

NASA administration decided on the dual-site option. To date, the Kennedy Space Center facilities have been established and are operating to take Shuttles into orbit. Vandenberg Air Force Base development has started and will become operational for Shuttle flights around 1985.

INAUGURATION OF REGULAR SERVICE

All of the orbital flight tests have been successful, and regular service began in November 1982. This first operational flight has initiated a series of 23 missions planned through 1985. A total of about 300 flights are scheduled between 1982 and 1995, the proposed lifetime of this first generation system. On the average, there will be two flights per month. Each flight requires extensive planning, training, and proper use of resources. We are about to step into 1985 and get a glimpse of flight preparations for a typical Shuttle mission.

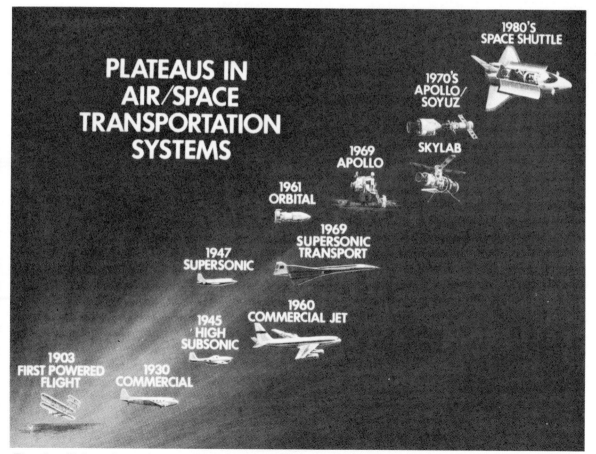

That first flight on December 17, 1903 was the beginning of a fabulous and fantastic 75-year period of innovation and invention, permitting man to fly in the air and in space. Beyond this, man's only limit is his imagination.

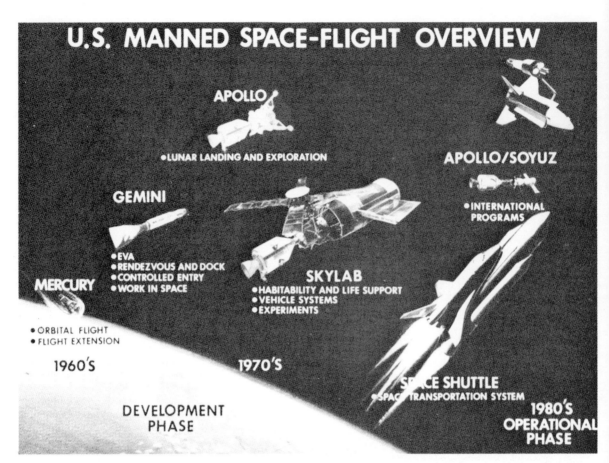

Every new endeavor has its skeptics. The usefulness of man in space is no exception. Many questioned the need for Mercury, Gemini, and Apollo flights. However, Skylab and the Shuttle have converted most skeptics to supporters of man's role in making space commonplace.

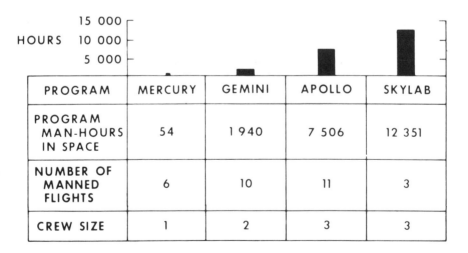

PROGRAM	MERCURY	GEMINI	APOLLO	SKYLAB
PROGRAM MAN-HOURS IN SPACE	54	1 940	7 506	12 351
NUMBER OF MANNED FLIGHTS	6	10	11	3
CREW SIZE	1	2	3	3

Pre-Shuttle cumulative staff hours in space 21,851 hours 25 minutes 41 seconds

CHAPTER TWO

LIFTOFF MINUS TWO WEEKS AND COUNTING

FREIGHTLINE OF THE 1980's

The time is 1985, and the Shuttle is flying to and from space on a regular basis. People in the United States and all over the world have accepted the Space Transportation System as a fact of life, and little excitement is aroused with the more or less biweekly launches and returns. The Kennedy Space Center will have been in operation as a U.S. shuttleport since 1982 and Vandenberg Air Force Base is just becoming operational. The Shuttle should indeed be an institution, just as the railroads were many years ago and as airfreight has more recently.

Let's look at airfreight service as we have known it in the 1960s and 1970s. This will serve as a good base from which to launch ourselves into future national spaceline operations. Think, for a moment, about airfreighting a package from New York City to Los Angeles. Assume we are sending a valuable piece of equipment. It would first be carefully prepared and packaged for shipment. Unless it were an animal, no provision is made for feeding or ventilation. Thus, a typical package is sealed and labelled. When it is fully prepared, you would take it to the airfreight counter at your local airport or have a trucking company take it from your plant to the airport. Once reaching the airfreight company, it is weighed, tagged, and scheduled for the next available freighter. All data on your

package are probably computerized to keep track of its location while en route. You will probably prepay for shipment or make some arrangement for delayed payment if you are a regular customer.

The plane arrives for loading. Shipments to Los Angeles are loaded, including yours. The position of each package has been computed to satisfy the weight and balance requirements of the airplane. This is critical to safe handling of the aircraft. The complete manifest list is printed out by a computer showing weight, size, destination, etc. of each piece of cargo. The flight crew arrives, usually consisting of a "salty" captain, a young co-pilot, and a flight engineer. Of course, no stewards or stewardesses are needed for the cargo. Actually these cargo flights are usually flown at night and are deadly dull. Crews have been known to catch up on their sleep along the way. In fact, on one such flight to Los Angeles a few years ago all three crew members fell asleep after engaging the autopilot. They were abruptly awakened by the yell of an air traffic controller over the radio receiver. Apparently they had flown right by Los Angeles and were a good part of the way to Hawaii.

We will assume your crew is wide awake. They meticulously check the reports about weather along the route, file a flight plan, and review the aircraft fuel load and cargo. If all checks out, they board the plane and prepare for departure. The captain calls Clearance Delivery over the radio for routing across the continent. Engines are started in proper sequence on the big four-engine jet. This is followed by a call to the ground controller for taxi instructions to the active runway. The controller gives them the runway information and taxi routing. Ground crew personnel pull the chocks and the engines are revved to taxi power. Your package has now begun its long journey to Los Angeles.

Upon reaching the active runway, the captain requests takeoff clearance from the tower. He gets it, and the huge jet turns onto the runway. All four engines are pushed to full thrust, and the airplane accelerates to rotation speed. Up, up, and away into the black sky for the 2,500-mile trip. "Gear up, trim for climb," commands the captain, as he adjusts power and maneuvers to satisfy noise abatement procedures.

The airplane quickly climbs to 38,000 feet for cruise at a speed of 600 miles per hour. Headwinds make the ground speed a little bit slower at 500 miles an hour. Activities on the flight deck settle down to checking gauges, filling in flight logs, and an occasional radio transmission with air traffic control. Four and one half hours later, the descent into the Los Angeles Basin begins. The crew again gets busy with controlling the aircraft's configuration, checking fuel levels, and stowing charts and other forms. Approach Control gives heading directions to the captain as he approaches the end of the landing runway. The plane touches down after a flight of five hours and fifteen minutes.

A well-lit freight hangar looms up ahead as

The Kennedy Space Center is located on approximately 140,000 miles of Merritt Island in central Florida. Most of the center buildings and launch complexes are depicted here, with the Shuttle facilities at the north end of the island. Two launch pads are available for the Space Transportation System, 39A and 39B. A 15,000-foot concrete runway will be the landing facility for returning Orbiters.

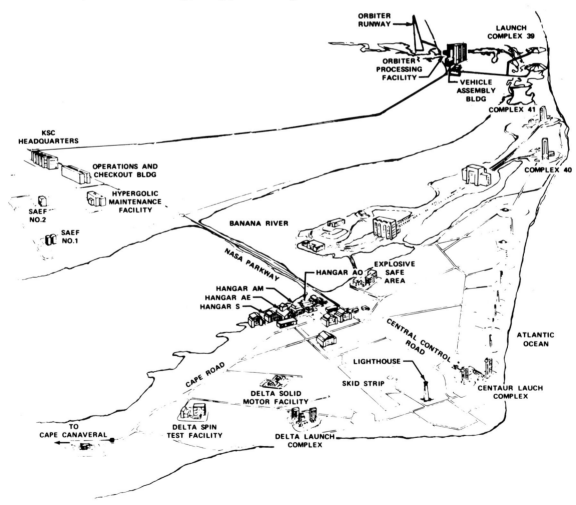

the plane taxies to the unloading area. Within two hours, the aircraft is unloaded, the crew files all paperwork, and local freight trucks arrive to deliver your cargo and the others' to their final destinations in the Los Angeles area. The trip ends as one of these trucks drops your package at the address on the shipping label. If the contents are equipment items for immediate use, then the person receiving the shipment will put them into service. Mission completed.

There are several fundamental differences between airfreighting a package between New York and Los Angeles and sending something into orbit. With expendable launch vehicles, there is no person receiving cargo in space. Most payloads are an end in themselves. They must be self-sustaining over their orbital lifetime, be able to communicate with Earth, and be maintenance free. Let's briefly look at the launch of weather satellites in the mid-1970s. A variety of standard boosters have been available over the years. However, the selection of a particular one for this type of satellite is quickly narrowed to one or two choices because of its size and orbital destination. Weather platforms in space have been launched from both Kennedy Space Center and Vandenberg Air Force Base, depending again on the final orbit to be achieved.

Since an artificial satellite is complete in itself, it must be built to last, maintenance free, and to work at high efficiency. These

marvels of modern technology take years to design and build. You just don't go into the company warehouse and take a handful of components off the shelves. A typical satellite requires tens and sometimes hundreds of custom-made components and structural members. They are incredibly complicated because of their mission in the remote cosmos. Early in the development of a new satellite, planning for its ultimate launch and operation in space must be begun. For weather satellites, decisions must be made about things like the radio frequencies of transmission, location of receiving stations on the ground, and the kinds of pictures to be taken from space.

Arrangements for transportation must also be made. In this case, transportation is an expendable booster. The arrangements consist of contacting the builder of the selected launch vehicle, as early as possible, and ordering the model and accessories that you want. Yes, this is something like buying a new car. After you select the make and model, then comes the hard part, whether to get air conditioning, AM-FM radio, and so on. Launch vehicles are something like that. There are different makes and models and a wide range of options that have to do with things like navigation, orbit shape, and radio transmissions.

At about the same time that you are ordering your launch vehicle, the National

The 15,000-foot concrete runway at the Kennedy Space Center is located near the Vehicle Assembly Building. Notice the drainage ditch around the runway for quick water runoff during heavy rains. This is similar to runways used by jet aircraft, except that it is somewhat longer. Typically, large airports have runways of between 10,000 and 12,000 feet.

Aeronautics and Space Administration should be contacted concerning the actual launch of your satellite and booster from either the east coast or west coast of the United States. NASA is responsible for all spacecraft launches in the United States. These activities reached a peak in 1966 with a total of 74 launches. During the 1970s, the U.S. has averaged about 30 launches per year. Arrangements should be made for your launch as soon as possible, as early as one or two years before the expected liftoff date. This is a wise action because there are only one or two launch pads available and limited personnel to handle the many tasks associated with the particular booster you are going to use. Remember, these facilities are used by all U.S. companies and government agencies sending spacecraft into orbit and to the planets. In addition, most foreign countries launch their spacecraft from our sites, with the exception of the USSR. Thus, there is usually a waiting list for any particular launch pad. In fact, there was a situation a few years ago when a foreign country contracted with a U.S. company to build a communications satellite to be used

solely for communicating between segments of their own country. All arrangements were made for the construction and designing of the satellite. However, the government officials involved neglected to arrange for a launch of their satellite. When the time came to ship the satellite to the Cape, NASA told them that they did not have a launch pad or a launch vehicle for them to use. Many *red* faces were observed in the space agency of that country. They had to delay the launch for almost two years while they made the arrangements for a booster and a launch date. Since you don't want your satellite sitting around waiting for a ride, you will not neglect to make these extremely important arrangements.

This brings us to America's new transportation system and into the 1980s. The Shuttle represents a delivery service to orbit, replacing all expendable boosters. Thus, there will be only one *make* of vehicle available. There will still be a wide variety of options to select from. This includes things like the number of crewmembers and their duties, launch site, orbit to be achieved, time of launch, and so on. All planning and arrange-

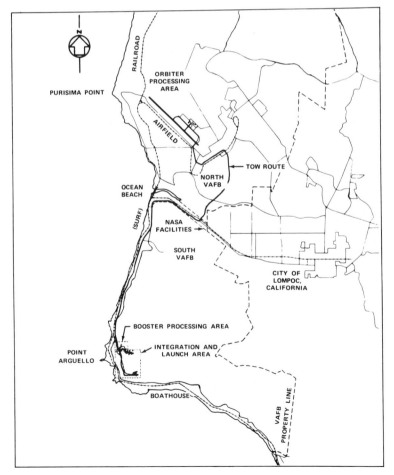

Vandenberg Air Force Base is scheduled to become the second operational launch site for the Space Shuttle in 1985. This facility is under the direction of the U.S. Air Force, which will be responsible for the Shuttle equipment and facilities necessary to perform missions from this base. It is located some 42 miles northwest of Santa Barbara, California. Many military and civilian spacecraft have been launched into orbits which allow them to fly over the polar regions of the world. Shuttle launches will take place from Space Launch Complex 6, which is located on a high plateau overlooking the ocean.

ments are made with NASA for the Shuttle, because this agency operates the National Spaceline. Schedules are made up far in advance, and the customer must reserve early. Let's imagine that you work for the National Weather Service and are sending up a satellite for weather pictures and other related measurements. The next few pages review the process you will go through for planning, paying, and preparing for your satellite launch on the Space Shuttle.

GROUND PROCESSING AT THE CAPE

For each orbital flight, planning begins years ahead of time with the design of the payload or special mission and the request for a particular flight assignment. All work and expense culminates in a launch at T minus 0, and your satellite or other device is on its way into space. Actual launch activities related to your particular Shuttle flight begin as late as two weeks before liftoff. This is the planned turn-around time for an Orbiter which has returned from a previous mission. Thus, the Shuttle and its support system are to be able to relaunch an Orbiter from the Kennedy Space Center within 160 working hours after it returns from a mission. These 160 hours are equivalent to ten working days with eight-hour shifts and two shifts per day in a five-day week. Based on this schedule, there are 14 calendar days in the turn-around period. This ground time has been compressed as much as possible at the Kennedy Space Center in the interest of decreasing both the maintenance expenses and the required number of Orbiters,

along with all the related equipment.

This huge spaceport is ideally suited to handle Shuttle landings and launches. Kennedy Space Center is located on approximately 140,000 acres on Merritt Island in Brevard County, Florida, and adjoins the Cape Canaveral Air Force Station. The elevation of the land is extremely low, with an average of only a dozen feet above sea level. This area is part of the Gulf Atlantic coastal flats. The site is situated on deposits overlying rock dating back to the Paleozoic Age. There are no caverns or significant metal or mineral deposits in the region, and there have been no earthquakes since October 1973. Before that, only rare, weak shakes were experienced. Soil on the island is generally sandy with the Banana and Indian Rivers lying to the east and west of Merritt Island, respectively. The local climate can be described as humid subtropical. Approximately 49 inches of rainfall is expected annually with fairly uniform distribution throughout the year. The winter temperature minimums are typically 55°F, and for summer, lows are 74°F with highs of 88°F. Thunderstorms are common in the spring and summer with an occasional severe storm. The Cape is subject to cyclic sea and land breezes in the summer. During spring, southerly and easterly winds prevail. Fall and winter winds are generally out of the north and east. There is an abundance of animals and birds in the region. Citrus fruit production is the predominant agricultural business in the area; there are several groves within the center boundaries that are leased to private

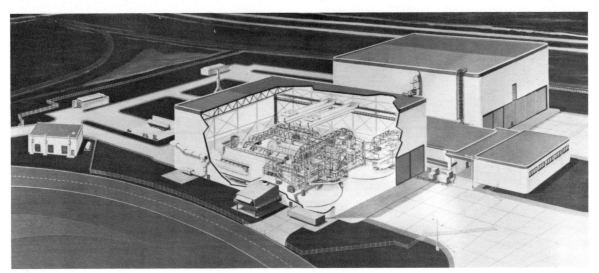

The Orbiter Processing Facility is a building in which payloads are removed from and inserted into the Orbiter, propellants are unloaded or loaded, and other parts of the Orbiter are refurbished.

Here we see the Orbiter Processing Facility nearing completion during the Spring of 1977. It is now receiving Orbiters and processing them. After completion of services each Orbiter is then taken to the Vehicle Assembly Building which is nearby.

individuals. Another activity associated with the citrus groves is a one-million-dollar honey industry. Under agreement with the U.S. Fish and Wildlife Service, the boundaries of the Merritt Island Wildlife Refuge and Kennedy Space Center coincide so that the Fish and Wildlife Service oversees all property not related to the space program.

Space launch complex 39, the site of the Apollo Lunar liftoffs, Skylab orbital missions, and the U.S. Apollo launch for the Apollo-Soyuz Test Project, can be used for up to 40 Shuttle launches per year from the eastern United States. Construction activities were primarily confined to areas previously used for industrial activities and had little or no adverse effect on the natural environment. The single most expensive construction in terms of land area was the 15,000-foot-long concrete runway and associated apron and tow strips. This was carefully designed and constructed to satisfy Shuttle needs. For example, a drainage ditch was included around the runway for quick water runoff during heavy rainfalls. However, one environmental aspect was overlooked. It turns out that alligators are extremely fond of such ditches and of sunbathing on warm concrete. Early test flights onto the new runway were preceded by jeeps driven up and down to chase off the sunbathers. Since this would eventually prove hazardous to landings as well as to alligators, a low fence was added around the runway to discourage such sunbathing.

Vandenberg Air Force Base is scheduled to become the second operational launch site for the Space Shuttle in 1985. Turn-around time is not so critical since only about twenty launches per year will occur from this base. It is located in Santa Barbara County, California, approximately 42 miles northwest of Santa Barbara and 120 miles from Los Angeles. The surface geology of the surrounding area is one associated with mountain building processes. This is confirmed by numerous fault zones and periodic earth-

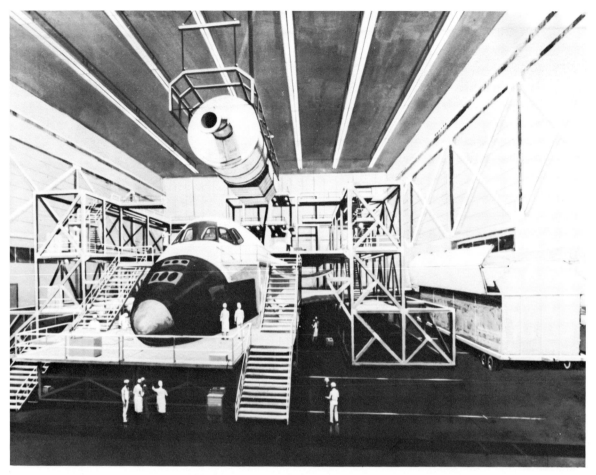

A Spacelab payload is carefully lowered on a "strongback" into the cargo bay of the Orbiter during preparation for a mission. This takes place in the Orbiter Processing Facility.

quakes. Nevertheless, the base has sustained no recorded quake damage. The southern California climate along the coast provides warm summers and usually mild winters. Generally, there is abundant sunshine and only a few rainy days. A well-known feature of this area is the persistent night and morning low cloudiness and fog, followed by sunny afternoons. Daytime winds are generally brisk and westerly; whereas, nighttime winds are rare. The total annual precipitation for this area is typically about thirteen inches.

Most areas of the Vandenberg Base where Shuttle activities are planned have already been developed into buildings or managed grasslands. The existing launch complex is known as Space Launch Complex 6 and is the one being modified for the Space Transportation System. It is centered on a plateau several hundred feet above the ocean. The complex is surrounded by a coastal scrub that is currently being grazed by domestic cattle. Animals are abundant in this region, with principal small species including California Valley quail and three types of rabbit. Mule deer and feral pigs are the major large animals. This area is also a known habitat for several threatened and endangered species. Of course, human species also live in the neighborhood, and extensive recreational facilities exist in the vicinity. For public safety, it is sometimes necessary to close one or more of the three nearby parks when missile launches are scheduled. It is anticipated that the period of closure will be no more than one day for Shuttle launches, for parks at both Vandenberg and the Cape.

Activities for your launch begin with an Orbiter landing on the 15,000-foot runway at the Kennedy Space Center. It takes about one hour to prepare for removal of the flight crew and the attachment of ground cooling and towing equipment. Since the wing and nose of the Shuttle are extremely hot from reentry into the atmosphere, it is necessary to pump cooling water through the wing structure as soon as possible after landing until the temperature drops to some safe zone. If this is not done, the reentry protection system will have to undergo more extensive refurbishment for the next flight. After cool-down, the Orbiter is towed to a place called the Orbiter Processing Facility where it is made safe for groundcrews to work on it. This involves the removal of propellants and explosive materials. Various special options from the previous flights are removed, and maintenance activities begin. Small propulsion devices are refurbished and reinstalled; the vehicle is checked out; the new payload, including your satellite is installed; and the connections between the Orbiter and payload items are checked. These activities consume about 96 hours of the allotted 160 hours. While the Orbiter is in this facility, the solid rocket boosters are stacked and aligned on the mobile launch platform in the famous Vehicle

Assembly Building where the Apollo vehicles were assembled several years ago. This is followed by placing the external propellant tank on these rocket boosters. The Orbiter is then moved to this building, lifted, rotated, on its end, and attached to the external tank. Finally, all connections are checked carefully. There are literally hundreds of electrical wires, pipes, and hoses running between the Orbiter and the external tank. The final activity in the vehicle assembly building is the installation of ordnance devices such as explosive bolts and separation devices. The time allotted for all this is 39 working hours. After this point, the Shuttle vehicle and the mobile platform can be moved or maintained for a long period of time. The move to the launch pad, making all connections, fueling, checkout, and launch take a minimum of 24 working hours. Certain payloads can also be installed right at the launch pad. This is par-

The Orbiter is boosted into a mating position with the External Tank and Solid Rocket Boosters in the Vehicle Assembly Building at Kennedy Space Center. This step occurs after completion of work in the Orbiter Processing Facility. Payloads may have already been placed in the cargo bay. After mating, the assembly is moved to the launch pad.

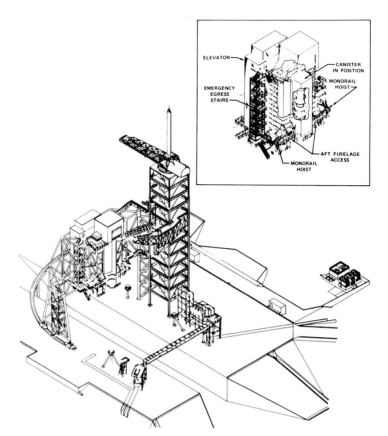

The Payload Changeout Room is a huge structure for handling payloads at the latest possible moment before launch. It is located right at the launch pad and is used for vertical installation of payloads into the cargo bay.

ticularly true for military satellites. The system is designed to be capable of launch within two hours after starting the filling of the rocket propellant tanks.

THE VEHICLE AND ITS COMPONENTS

It should be evident by now, that the Space Shuttle System is composed of four major components: Orbiter, External Tank, and two Solid Rocket Boosters. Since the crew and cargo are housed in the Orbiter, it has received the greatest public exposure and attention. In fact, this is the only component of the major four which actually gets into orbit. The Solid Rocket Boosters function for only two minutes at the very beginning of each flight. The External Tank carries propellant for the three Shuttle main engines from liftoff to an altitude just short of actual orbital flight. The Orbiter and boosters are reusable, but the External Tank is expended on each launch. As late as 1972, several basic Shuttle configurations were still being considered. These included systems with flyable boosters, tandem stacking, and various piggyback arrangements, but the external tank and recoverable rocket booster concept prevailed.

Normally, a crew of three or four and a payload are carried into orbit for up to seven days. For some special experiment payloads, flights of up to 30 days can be accomplished. Also, provisions are available for up to seven crewmembers, some of whom could work in experiment modules such as Spacelab, a set of European built laboratories which remain bolted in the cargo bay for each of its flights. The Shuttle could also be used to rescue other crews in orbit with just a 24-hour notice.

The solid boosters burn in parallel with the main rocket engines on the Orbiter. They separate from the vehicle at an altitude of about 150,000 feet and descend on parachutes into the ocean about 140 miles from the launch site. Tugboats are in the vicinity to complete this recovery maneuver. They tow the boosters back to a refurbishment facility and for sequencing into another launch. The external tank is the only source of powerful liquid oxygen and liquid hydrogen propellants for the three main rocket engines. Thus, it is carried almost all the way to orbit. Just seconds before achieving orbital velocity, it is separated and makes a ballistic type reentry back into the atmosphere which largely destroys it due to aerodynamic heating and stresses as it descends. On a typical launch

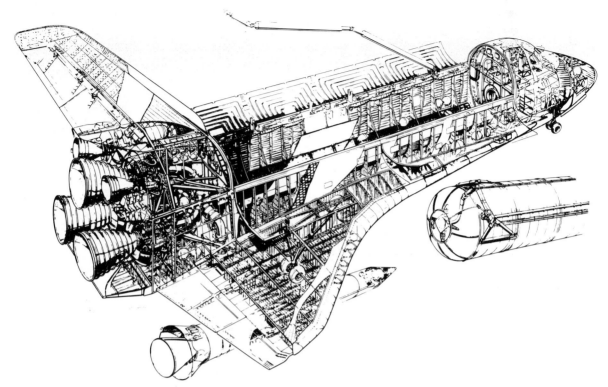

This cutaway view of the Orbiter gives a good idea of the complexity of the system. The three large nozzles at the rear of the Orbiter are part of the main rocket engines, while the two smaller nozzles are part of the Orbital Maneuvering Subsystem. Structurally, this vehicle is somewhat like an aircraft. The only pressurized component is the crewcabin in the forward section.

Several configurations of the Space Shuttle were being considered in the early 1970's. A Saturn V is shown here for comparison with the contenders. From left to right, we see a Grumman configuration which uses a reusable Orbiter with external hydrogen tanks mounted on a stage of a Saturn booster. Next is a McDonnell Douglas reusable Orbiter and booster. The third one is a Grumman Orbiter with external hydrogen tanks and reusable booster. Finally, a Rockwell configuration shows a reusable Orbiter and booster. As we know now, none of these configurations was selected as the final one.

Workmen at Thiokol Corporation's Division in Utah are shown here applying insulation to the inside of a Solid Rocket Booster casing. This should give you a good feel for the size of these boosters.

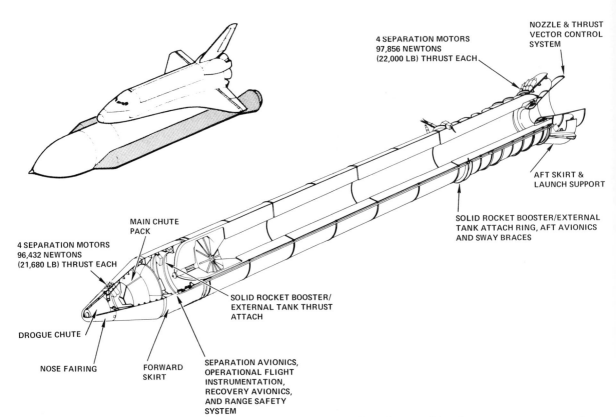

NOZZLE & THRUST VECTOR CONTROL SYSTEM

4 SEPARATION MOTORS 97,856 NEWTONS (22,000 LB) THRUST EACH

AFT SKIRT & LAUNCH SUPPORT

SOLID ROCKET BOOSTER/EXTERNAL TANK ATTACH RING, AFT AVIONICS AND SWAY BRACES

MAIN CHUTE PACK

4 SEPARATION MOTORS 96,432 NEWTONS (21,680 LB) THRUST EACH

SOLID ROCKET BOOSTER/ EXTERNAL TANK THRUST ATTACH

DROGUE CHUTE

NOSE FAIRING

FORWARD SKIRT

SEPARATION AVIONICS, OPERATIONAL FLIGHT INSTRUMENTATION, RECOVERY AVIONICS, AND RANGE SAFETY SYSTEM

Two Solid Rocket Boosters are used during each Shuttle launch. These huge solid rocket motors are over twelve feet in diameter and 149 feet in length. Each weighs about 1.3 million pounds and produces about 2.9 million pounds of thrust at liftoff. They burn for only two minutes and are then separated from the External Tank. Parachutes bring them down slowly into the ocean where they are recovered and later reused. On STS-4 neither parachute worked. Both boosters impacted the water so hard they sank to the ocean bottom. NASA sent the Navy out to look for them.

the resulting debris will fall into ocean areas approximately 11,500 miles from the launch site, halfway around the world. The Orbiter continues on into orbit by firing a pair of small rockets used for maneuvering in space.

Each Orbiter is designed for use on 100 flights with some refurbishment after each one. The accompanying solid rocket booster looks simple even though it is so large. It is actually quite complicated and consists of a segmented case, solid propellant, igniter, nozzle, separation system, recovery electronics, parachutes and range-safety destruct system. The case is designed to be used 20 times. The external tank is an expendable component and is designed for a single flight at minimum cost. It is, nevertheless, sophisticated because of its key function of providing all fuel and oxidizer at prescribed pressures, temperatures, and flow rates to the three main rockets. This large, cigar-shaped container is divided into two separate holding tanks since the fuel and oxidizer must not come into contact until reaching the rocket motors on the Orbiter.

The Orbiter is indeed a strange looking vehicle. One must really classify it as part air-

craft, part spacecraft. The single most important factor that distinguishes this vehicle from other space-bound carriers is that it returns and is reused over and over again. Furthermore, it does this while carrying huge payloads and a crew at an economical price. Sizewise, the Orbiter is comparable to a McDonnell Douglas DC-9; they both have about the same length, with the Orbiter having a shorter wingspan. Both birds are composed of several complicated subsystems and components. One of the most important Orbiter ingredients is the main propulsion system. It consists of three rocket engines located in the aft fuselage area, the external tank, plumbing, and controls. Another critical thrust-producing part of the Orbiter is the Orbital Maneuvering Subsystem, positioned in two unaerodynamic looking pods above the main rockets. These units are used for orbit insertion and to provide for small orbit changes, rendezvous, and de-orbit.

The main propulsion system operates for about eight minutes, from just before liftoff until main engine cutoff. Each engine produces a combustion chamber pressure of

3,000 psi, about 204 times that of the atmosphere at sea level. This represents a significant new technological advancement. A key in producing thrust is the chamber pressure; the higher it is, the higher the thrust. This is the highest pressure yet used in an engine of this size. Each main rocket can be swiveled to provide control of the Orbiter orientation during ascent.

The Orbital Maneuvering Subsystem has its own propellant tanks and can be removed and worked on separately from the Orbiter. The built-in tanks carry just enough propellant to give the Orbiter that last little kick into orbit as well as to provide for a small amount of maneuvering capability and de-orbit thrusts. Many missions will require more maneuvers, leading to the use of propellant kits to augment the built-in supply. Up to three such kits can be installed in the cargo bay. Unfortunately, the weight and volume of these kits must be subtracted from maximum payload capacities of the system.

GETTING THE PAYLOAD READY

While the various components of the Shuttle vehicle are being refurbished and assembled, new and overhauled payloads will be assembled and tested elsewhere at the launch site. These will be transported to the Cape by means acceptable to the customer. Payloads may arrive at the Kennedy Space Center by air, overland or water. One method of overland shipment that has been acceptable to many satellite programs in the past is by means of a commercial air-ride van provided with environmental controls. The payload is usually mounted on a platform and protected by a soft cover. It is estimated that this system could accommodate most spacecraft being flown on the Shuttle. This sounds simple enough; however, don't for a moment assume that the overland transportation mode is a simple matter. It's not like getting a U-haul trailer and hooking it to the back of your car. Electrical power, air conditioning, and a smooth ride must be provided to most payloads. NASA will, however, supply the transporter to your facility on a pre-selected date. It is then your responsibility to bring the payload and transporter to the Cape.

Most payloads will be installed horizontally while the Orbiter is in the processing facility. Placement of the various cargo items is carefully planned to insure proper balance of the load. The Shuttle, like all aircraft, must

One External Tank is used on each Shuttle launch. It is 27.5 feet in diameter and 154 feet long and weighs 1.6 million pounds at liftoff. This tank provides all of the propellants to the three Orbiter main engines which burn from liftoff to just before reaching orbit. The External Tank is then separated and reenters the atmosphere. Most of it is burned up during reentry, with the remaining pieces falling into the Indian Ocean.

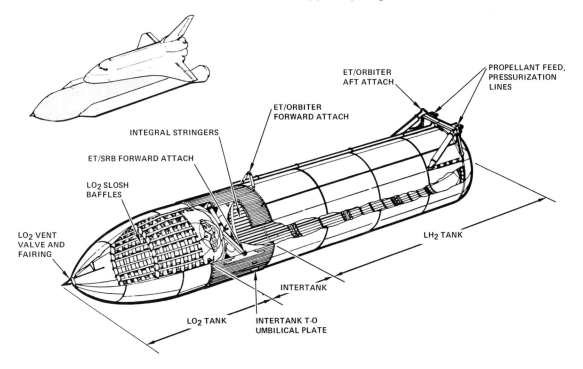

ORBITAL MANEUVERING SYSTEM

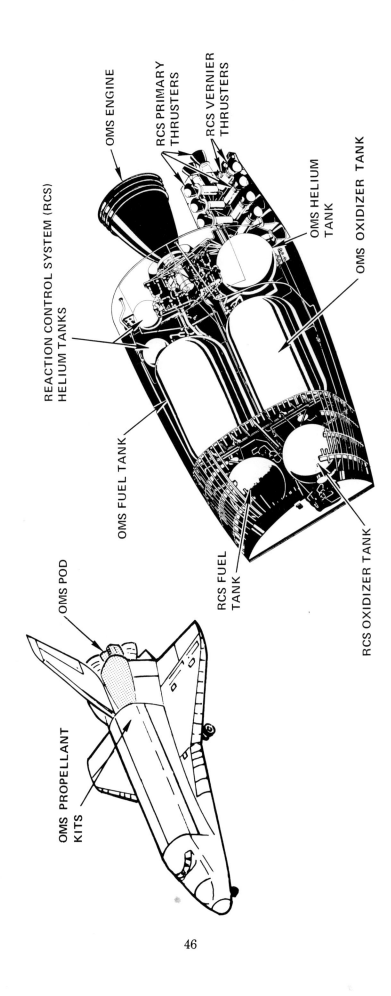

OMS ENGINE

RCS PRIMARY THRUSTERS

RCS VERNIER THRUSTERS

REACTION CONTROL SYSTEM (RCS) HELIUM TANKS

OMS HELIUM TANK

OMS OXIDIZER TANK

OMS FUEL TANK

RCS FUEL TANK

RCS OXIDIZER TANK

OMS POD

OMS PROPELLANT KITS

The Orbital Maneuvering Subsystem (OMS) consists of a set of two rocket engines, each producing 6,000 pounds of thrust in orbit. They are located in a pair of pods, one on each side of the vertical stabilizer. This system has its own basic set of propellant tanks, but additional sets of tanks can be added to the cargo bay if needed for a specific mission.

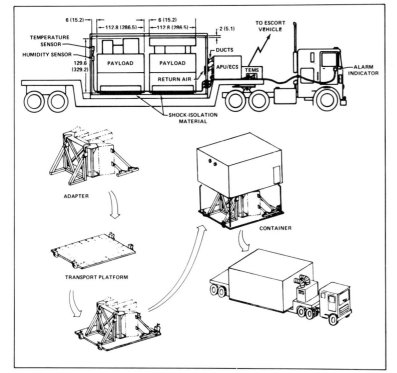

Here is a schematic showing the elements of the standard transportation system and the type of commercial carrier to be used in bringing payloads to the Cape. The container has its own environmental control system, with power being provided by an auxiliary power unit. A battery is included to supply power to the payload and operate the environmental monitoring system for at least four hours if the generator should fail. Larger payloads, such as Spacelab, require a somewhat different type of payload container.

satisfy strict weight and balance limitations for proper control when flying through the atmosphere. If your flight includes a Spacelab, it already has its experiments on-board when loaded into the Shuttle. Military payloads and a few others require vertical installation. These are brought to the launch pad in an environmentally controlled canister. They are ready for installation and are put into the Orbiter through the use of the Payload Changeout Room, the huge structure for handling payloads at the latest possible moment before launch. For example, a combination spacecraft and upper stage will be mounted in the Orbiter at this point.

After the payload bay doors are closed, the internal environment will be maintained from the processing facility to the Vehicle Assembly Building and then onto the launch pad. This will provide for protection against contamination of sensitive instruments due to dust, smoke, chemicals and heat. However, there will be a period of approximately 40 hours during Orbiter hoisting operations in the Vehicle Assembly Building when environmental control will be interrupted. This should not be critical since the cargo will be sealed at that time. While the payloads are in the Vehicle Assembly Building, they are more or less left alone, with the exception of monitoring of potentially hazardous systems. Electrical power is not available to the cargo

during towing to the Vehicle Assembly Building, Orbiter assembly and mating, and transfer to the launch pad. Thus, the payloads will have to be capable of withstanding this relatively short period of time on their own. After all the components of the system

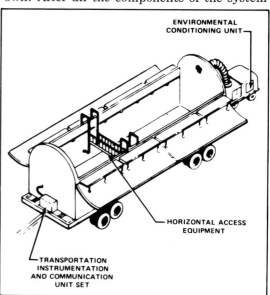

Large payloads, such as Spacelab, require a long transporter. Here is a schematic of such a device which also includes an environmental conditioning unit. A crane is used to lower Spacelab into the payload bay during preparations for flight. A special horizontal handling fixture called a "strongback" is used to support the long structure while being loaded.

are assembled and checked in the Vehicle Assembly Building, the entire system is transferred to the launch pad on the Mobile Launch Platform. Payloads that require vertical installation are moved to the launch pad in the environmentally controlled canister which is the same size as the Orbiter cargo bay. These are installed by a huge handling mechanism inside the Payload Changeout Room. Cargo doors are again closed, and extensive checks are made involving both the Shuttle components and its precious load. Countdown preparations continue until two hours before launch. At that time, the supercold liquid oxygen and liquid hydrogen propellants are loaded, the flight crew boards, and final countdown is begun.

THE FARE FOR CARGO

Don't expect regular passenger service to orbit and back for several years to come. You cannot buy a ticket for yourself, but you can purchase one for payloads such as satellites and experiments. Cargo is the main concern of the Space Transportation System, at least un-

til the early 1990s. Accordingly, a fare schedule has been developed by NASA. This schedule reflected a policy intended to encourage maximum use of the system by operating in the *red* initially. However, NASA plans to more than double this price structure in 1985. Users will be charged as much as $98 million to launch the Shuttle each time. This may go up to $106 million in 1986. In addition to this, there will be a $4.2 million fee for each customer on each flight. The price per launch between 1982 and 1985 is a mere $42 million, including the $4.2 million fee. Fortunately, most Shuttle flights carry an average of three satellites allowing the launch charges to be prorated according to volume in the cargo bay.

These price increases result from a combination of inflation, reduced number of missions, unexpected cost increases in hardware, and greater-than-expected refurbishment costs between flights. The increased hardware costs are largely related to the production of the solid rocket boosters and a lightweight external tank, 8,000 pounds lighter than earlier tanks. Other sources of increased cost

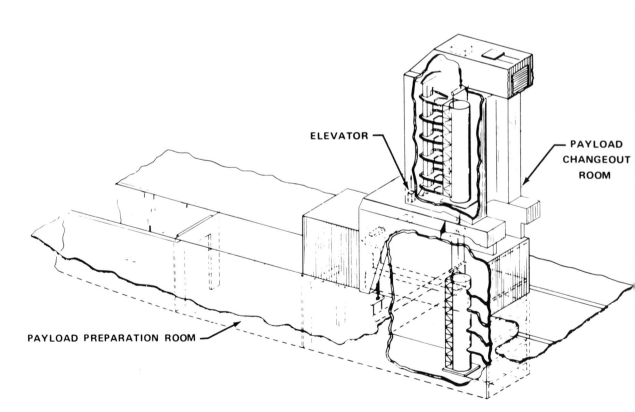

ELEVATOR

PAYLOAD CHANGEOUT ROOM

PAYLOAD PREPARATION ROOM

At the Vandenberg Air Force Base, Shuttle payloads will be loaded either at a facility very similar to the Orbiter Processing Facility or at the launch pad by a payload preparation and changeout room. The payload preparation area at the pad is underground. Spacecraft are made ready to be loaded and then lifted in a canister up to the launch area. They are then inserted into the payload bay in a vertical position.

48

include hardware transportation to Florida and additional staff needed at the launch site.

The actual processes of securing a "berth" and of computing the cost are extremely complicated. For example, NASA has decided to quote prices in terms of the value of the dollar in 1975, so the price given to you from the NASA Shuttle User office will appear lower than those listed above. A number of fees and options also come into play. In reviewing the price schedule, I had the same feeling that I usually get when trying to purchase an airline ticket in the least expensive manner. There always seems to be an uncountable number of options with an infinite number of conditions attached. For example, the price charged non-U.S. Government users for Shuttle flights, whether domestic or foreign, is designed to cover a fair share of both the total operating costs and NASA's investment in the fleet, facilities, and equipment. Charges to civilian agencies of the U.S. Government and participating foreign government users are intended to cover only a fair share of the total operating costs. The Department of Defense has a unique situation. Since NASA and the military have provided each other with accommodations at their respective launch sites, defense satellites get a special rate.

Users with an exceptional new use of space or a first-time application of great value to the public are considered in a separate classification. The price charged for such a flight will be just the incremental cost of conducting one additional mission during the program. If you think your satellite may qualify for such a fare, there is an established process which is used to select such payloads. The final deci-sion is up to the NASA Administrator.

All prospective major users, regardless of class, must make a down-payment of $100,000 to NASA as *earnest* money before beginning negotiations for a flight. In other words, before they will even talk to you about flying your satellite, they want 100 "big ones." Furthermore, this is a non-refundable payment, but it will be applied to your total mission costs, if all goes well. Remember, the basic price does not include accessories such as the use of a Spacelab, upper stages for achieving higher orbits after Shuttle release, or other special equipment or services.

The basic billing schedule for all users begins 33 months before the planned launch date. If you can't plan that far in advance, it will cost extra, and payments will be made on an accelerated schedule. Really poor planners who want to launch within one year of negotiating for a flight assignment will be handled on a space-available basis or as a short-term callup option. For example, if NASA can accommodate a user who needs the full capacity of the Orbiter, the additional cost will be a penalty of up to 22 percent, depending on the time available for the accelerated launch schedule. The payment schedule is shown below with each payment expressed as a percentage of a nominal total price for a 33-month schedule.

In addition to the basic charges presented above, there are three price and schedule options available for a fee. One: users wishing to launch a payload after October 1, 1983 can contract with NASA now for a firm price. However, the security of fixing the price will cost an extra $1,000,000 of the "1975-kind."

Payment schedule for cargo fares on the Space Transportation System.
Each payment is a percentage of a nominal total price for a 33-month schedule.

Contract initiation in months before launch	Months before launch						Total (percent)
	33	27	21	15	9	3	
Nominal schedule (more than 33 months before launch site)	10	10	17	17	23	23	100
Accelerated schedule							
27 to 32		21	17	17	23	23	101
21 to 26			40	17	23	23	103
15 to 20				61	23	23	107
9 to 14					90	23	113
3 to 8						122	122

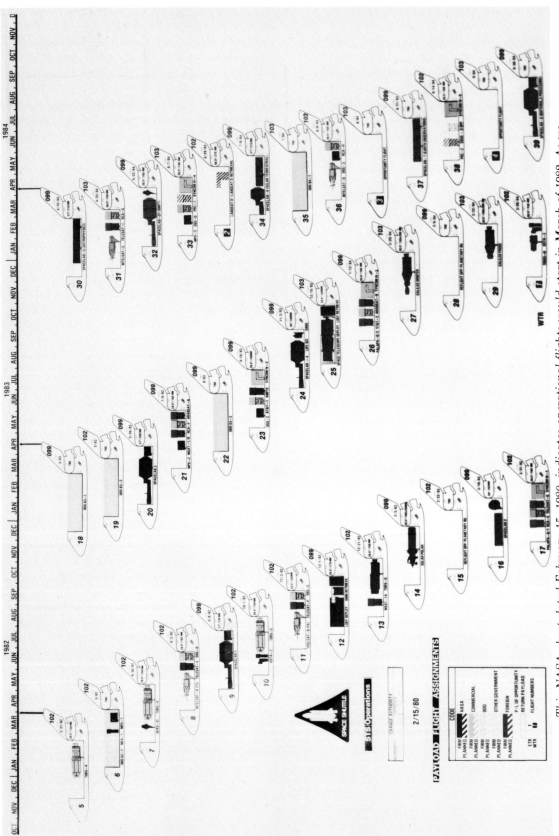

This NASA chart dated February 15, 1980, indicates operational flights would start in March of 1982. As we know, these flights began in the Fall of 1982. Thirty-nine such missions are depicted in which military, commercial, and civilian payloads are to be launched. Three different orbiters are used, 099, 102, and 103. Both Shuttle launch sites are employed, ETR and WTR.

50

This is payable at the time the $100,000 in earnest money is paid. If this option is selected, the cost will be the price of a flight on September 30, 1983 plus eight percent compounded annually from October 1, 1983 to the planned date of launch. Two: users can contract for a guaranteed launch date within a specified 90-day period by paying an additional fee of $100,000, payable at the time the $100,000 in earnest money is paid. Three: the *floating launch date* affords the user some flexibility in deciding when to launch. A date is tentatively selected at least 33 months away. Then, when the user notifies NASA of a desired launch date, a firm schedule is worked out. This option costs an extra ten percent of the flight price in effect when the contract is signed, and the fee must be paid at that time. If the user requests a launch date less than twelve months in advance, *short-term callup fees* also apply. This option permits the poor planner and other users who cannot make firm schedules to contract for a flight without specifying the launch date. Thus, no postponement fees would result, as they might without using this option.

All these options require that a contract be signed at least three years before the launch. Users can postpone a flight once at no additional cost if NASA is notified more than one year before the scheduled launch date. However, further postponements occurring less than one year before the planned launch will cost five percent of the Shuttle fare. Any time the user cancels a flight, the cost is ten percent of the fare. If postponement causes a payload to be launched in a year when a higher fare is in effect, the new fare will apply.

The basic charge schedule and option fees presented above are for a dedicated (single user) Shuttle flight. Other services available at an additional charge include revisit and retrieval, use of Spacelab or other special equipment, use of add-on supplies to extend the basic Shuttle on-orbit time, use of upper stages, extravehicular activities (EVA) by the crew, and possible special training. If your payload will not fill up the cargo bay, then the flight might be shared with another satellite (or satellites) going into a similar orbit. Then your cost will be a fraction of the dedicated flight price, calculated in a four-step manner. First, the payload weight is divided by the Shuttle payload weight capacity for the desired orbit to find the *weight-load factor*. The Shuttle can carry up to 65,000 lbs. when

launched in a due east direction from the Kennedy Space Center. This is reduced to 37,000 lbs. for a polar launch from the Vandenberg Air Force Base. Second, the payload length is divided by the length of the cargo bay, 60 feet, to find the *length-load factor*. For example, if your payload is 30 feet long, then the length-load factor is just one-half. Third, the *load factor* (length or weight, whichever is greater) is multiplied by 1-1/3 to determine the cost factor. For example, if you end up with a weight-load factor of 1/5 and a length-load factor of 1/4, then you would use 1/4 as the load-factor which is multiplied by 1-1/3. This gives you a cost factor of 1/3. Fourth, the calculated cost factor is multiplied by the price of a *dedicated* flight for your class of user. Thus, if the cost factor is 1/3, then it would cost you 1/3 of the dedicated flight price to fly your payload. It all sounds very complicated, but it is not. To demonstrate the use of this method, let's try another example. Assume your satellite weighs 10,000 lbs. when loaded into the Shuttle. The National Weather Service is a U.S. civilian government agency. If the spacecraft is compact, the weight will determine the launch fee. If the orbit you are going into is associated with a 65,000-lb. capacity for the Orbiter, then the weight factor is 10,000/65,000 or about fifteen percent of the capacity for the Shuttle. This gives a cost factor of 21 percent. If the flight costs a total of $42 million, then your part is about $8.8 million.

Now that you understand the basic pricing scheme, other factors affecting cost can be considered. A twenty percent discount will be given to shared flight users who agree to fly on a *space-available* or *stand-by* basis. Furthermore, small self-contained research-type payloads that require no Shuttle services such as electrical power, mechanical manipulation, or cooling will also be flown on a space-available basis, provided they weigh less than two hundred pounds and occupy less than five cubic feet of volume. The fare for these *getaway specials* will not exceed $10,000 (in 1975 dollars) and will be a minimum $3,000 with the exact amount depending on size and weight. Earnest money for this type payload is only $500. That's right, for only $500 you can reserve space for your own getaway special, provided it is not intended for commercial promotion purposes. In fact, NASA has received earnest money for over 200 such payloads from a diverse group of companies, universities, and individuals. The procedures for reserving space for a getaway special are straightforward. Your first step is to call NASA at (202) 755-2427 for information. You will be requested to submit $500 earnest money with a letter stating that you wish to reserve space for a small self-contained payload to:

Director, Financial Management
Code BF-6
NASA Headquarters
Washington, DC 20546

with a copy also to:

Director, Space Transportation Utilization
Code OT-6
NASA Headquarters
Washington, DC 20546

This will result in an acknowledgement letter assigning you a payload number. Technical discussions will then begin with the Goddard Space Flight Center.

For users requiring a major proportion of Shuttle payload space, additional decisions are required concerning the type of payload and launch operations which are most appropriate to your needs. There are certain terms and conditions imposed on both the user and NASA. For example, a non-U.S. Government customer is guaranteed a reflight. This assurance is included in the fee. The guarantee applies if the first launch attempt is unsuccessful through no fault of the user and if the payload returns safely to Earth or if a second payload is provided by the user. In the case of a retrieval mission, the launch of a Shuttle into the proper payload orbit is guaranteed if the first retrieval attempt is unsuccessful. In all instances, the fare does not imply that damage caused to the payload through the fault of the U.S. Government or its contractors will be compensated. The U.S. Government, therefore, will assume no risk for damage or loss of the user's payload. The customer can either take the chance of losing his payload or obtain insurance protection.

Many people have thought of bringing old satellites back to Earth. One organization which is particularly interested in doing this is the Smithsonian Institution. Such a feat is possible if it can be done safely. Many retrievals will, in fact, be performed to recover Shuttle launched satellites in the 1980s and beyond. These satellites, however, will be *designed* for retrieval. Satellites now in orbit

have not been so designed and may cause some problems when attempts are made to bring them back. NASA is willing to provide revisit and retrieval services on the basis of estimated costs. If a dedicated Shuttle flight is required, the full fare will be charged. If the user's retrieval requirement can be satisfied in connection with an already scheduled flight, the fee charged will be only for added flight planning, special hardware, and any other costs incurred by the revisit.

FLIGHT PLANNING

In addition to making arrangements for your flight, negotiating a fee and schedule contract, and preparing your payload, you, as the Space Transportation System user, must also devise a *mission plan* for your flight. NASA, of course, will work with you on planning, which starts well before the two-week pre-launch countdown, and it involves a variety of topics. More complex payloads will need more time for the planning process, but the basic time period is sixteen weeks for relatively simple flights.

There are four major items of concern in planning your mission. To begin with, the flight sequence must be worked out with excruciating care to make sure all activities are included. The exact trajectory of launch, liftoff time, and final orbit have to be specified. For example, you must know exactly which ground stations are needed to track the Shuttle and its payload. Where are the sun and moon at the instant of liftoff? When do the cargo bay doors open? What signals are being sent by your satellite to the ground? These questions are typical of the flight sequence planning needed to send your payload up.

Another major item is planning the flight crew activities. This involves specifying the exact duties of each person on your flight. For example, one crewmember will be assigned the tasks of opening the doors, getting your satellite ready for release, and then releasing it into space. Another person will be making sure that the Orbiter is properly oriented relative to the earth and sun. Meanwhile, there are hundreds of people on the ground, at various facilities around the world, who are working on getting your satellite into its final orbit and operating properly. Thus, the third major item for planning is concerned with the duties and responsibilities of these people. During liftoff and ascent, ground crews at the

STS 100 FORM	REQUEST FOR FLIGHT ASSIGNMENT	DATE:

To: SPACE TRANSPORTATION SYSTEMS OPERATIONS
MAIL CODE MO
NATIONAL AERONAUTICS AND SPACE ADMINISTRATION
WASHINGTON D.C. 20546

FROM: _____
_____
_____
_____

/S/

FLIGHT OBJECTIVES:

____ Earnest money   ____ NASA approved

____ Commercial   ____ ESA approved

____ Other Government ____ DOD approved

Flight period _____ or specific date _____ FLIGHT TYPE:

Inclination range _____ or specific inclination _____ ____ Deployment

Altitude range _____ or specific altitude _____ ____ Attached

Payload configuration _____ ____ Servicing

Flight duration, hours attached _____ Discipline _____ ____ Retrieval

Crew complement: Commander, pilot, mission specialist plus option for additional mission

    specialist(s) _____ or payload specialist(s) _____

Payload Operations Control Center support:

_____ GSFC _____ JPL _____ JSC _____ Other _____ Not required _____

STDN and Tracking and Data Relay Satellite system support (comment):

Payload mass properties including flight kits: Specify flight kits used in weight:

                                                      (see JSC-07700, Volume XIV)

    Weight: Launch _____ lb. _____ kg

               Landing _____ lb. _____ kg

    Diameter: Launch _____ inches _____ mm

               Landing _____ inches _____ mm

    Length: Launch _____ inches _____ mm

               Landing _____ inches _____ mm

Payload kWh estimate _____ kWh

Payload constraints and/or unique requirements:

Orientation, pointing, sunlight constraints, etc. (comment):

Special prelaunch and postlanding off-line support at launch and landing site (comment):

Special prelaunch and postlanding on-line support while in the Orbiter (comment):

If you have a payload to be loaded onto the Space Shuttle you must first get a firm flight assignment from NASA. The necessary form to begin the sequence to your launch is STS 100. Here is the front side of the form, which includes most of the pertinent questions. When this is received along with your earnest money at NASA Headquarters, the process of mating your launch requirements with a flight begins.

A view inside the mock-up of Spacelab gives us an idea of where the different equipment and instrumentation panels are located.

Kennedy Space Center in Florida and the Johnson Space Center (home of the Mission Control Center) in Houston, Texas, are watching close to 100 or more video screens (like TV monitors) which display thousands of bits of data, indicating the status of the Shuttle and its components. Everything from the pilot's pulse rate to the temperature of the cargo is instantly available to these flight controllers. Once into orbit, all important measurements continue to be monitored. Men and women at several NASA centers and at the user's facilities will be busily checking on every aspect of flight progress. The duties of every person involved must be carefully spelled out well before liftoff.

The final major item of planning is related to personnel duties: training and preparation for launch activities. When your Shuttle leaves the pad, every person working in the mission, whether on the ground or in the Orbiter, had better know his or her job and the

limits of responsibility associated with it. That is the reason for the training before each flight. Although most launches will have similarities, each has at least a few unique aspects. This training will guarantee that these are taken care of, and every activity goes smoothly. Eight weeks before launch, the flight crew and ground control team are selected for your mission, allowing plenty of time for training.

All qualified flight crewmembers have, of course, gone through extensive training for a number of years. They have used very sophisticated equipment at NASA facilities and at contractor plants such as Rockwell International. For example, the Orbiter One-g Trainer is a full-scale representation of the crew cabin and part of the cargo bay. Details, such as payload attachment points, are provided in the cargo bay area. This is used for crew training in the several functions required on orbit. There is also an Orbiter Neutral

The Spacelab Simulator consists of a cabin and an experiment segment. Teams of experimenters are being trained at this facility in Germany.

Buoyancy Trainer, used underwater to simulate weightlessness. It is a full-scale representation of part of the crew cabin, including airlock and cargo bay doors. This is especially useful for training in extravehicular activity procedures. The Shuttle Mission Simulator provides realistic generation of flight dynamics experienced in orbit. The Remote Manipulator System Trainer consists of an aft crew station mock-up, a cargo bay mock-up, and a mechanically operated arm. It allows the crew to gain experience in payload grappling and stowing. This should be a lot of fun, because each user of the Space Transportation System has to provide helium inflatable models to simulate the payload geometry. If the crew can grapple these lighter-than-air models, they can certainly handle any payload in space. Finally, the Spacelab Simulator consists of a cabin and an experiment segment, and is used to train teams which will work in this facility while in orbit.

One of the first planning steps for your flight is getting a firm flight assignment from NASA. Only a few simple steps are required in initiating a request and finalizing this assign-

ment. The necessary form to begin the long process to your Shuttle launch is called STS 100, "Request for Flight Assignment." It is a simple looking, two-sided form which asks for basic information like flight objectives, launch date, desired orbital altitude, payload weight and size, and so on. When this is received, along with your earnest money of $100,000 for a normal category user or less for a getaway special, at the Space Transportation System Operations office at NASA headquarters in Washington, D.C., the process of mating your launch requirements with a flight begins. A whole series of meetings and phone conversations with NASA staff members will take place over a period of weeks. At this point, a tentative flight assignment is made, and you have officially begun the trip to launch complex 39 some three years away.

CREW ACCOMMODATIONS AND DUTIES

The basic Shuttle vehicle components have already been identified, but the crew compartments have yet to be described. This part of the Orbiter is the center of all command and control in space. The cabin module is designed

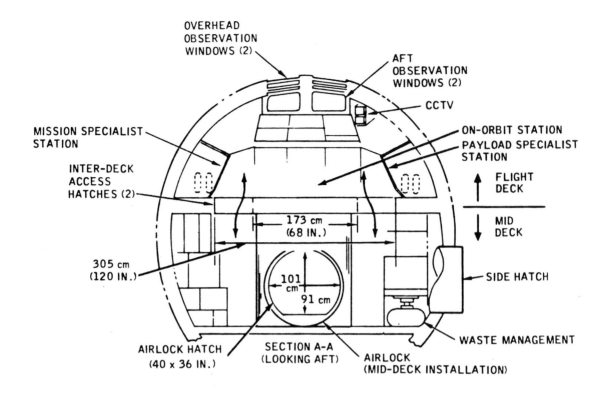

OVERHEAD OBSERVATION WINDOWS (2)

AFT OBSERVATION WINDOWS (2)

CCTV

MISSION SPECIALIST STATION

ON-ORBIT STATION PAYLOAD SPECIALIST STATION

INTER-DECK ACCESS HATCHES (2)

FLIGHT DECK

MID DECK

173 cm (68 IN.)

305 cm (120 IN.)

101 cm

91 cm

SIDE HATCH

WASTE MANAGEMENT

AIRLOCK HATCH (40 x 36 IN.)

SECTION A-A (LOOKING AFT)

AIRLOCK (MID-DECK INSTALLATION)

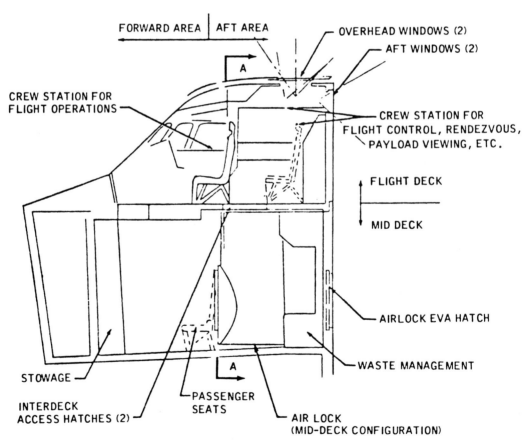

FORWARD AREA | AFT AREA

OVERHEAD WINDOWS (2)

AFT WINDOWS (2)

A

CREW STATION FOR FLIGHT OPERATIONS

CREW STATION FOR FLIGHT CONTROL, RENDEZVOUS, PAYLOAD VIEWING, ETC.

FLIGHT DECK

MID DECK

AIRLOCK EVA HATCH

A

WASTE MANAGEMENT

STOWAGE

INTERDECK ACCESS HATCHES (2)

PASSENGER SEATS

AIR LOCK (MID-DECK CONFIGURATION)

The crew cabin has three levels. Only the upper two are sketched here. The flight deck contains all displays and controls for flight. Provisions and the galley are located on the mid-deck.

A cutaway of the Space Shuttle vehicle shows the important subsystems and components. The cargo bay is 15 feet in diameter and 60 feet long, with the crew cabin immediately in front of this bay. The Orbital Maneuvering Subsystem thrusters have their own propellant tanks neatly located near the engines themselves. The External Tank is really two tanks, one for liquid oxygen and one for liquid hydrogen. The Solid Rocket Boosters are primarily casings for huge solid propellants. In addition, they have a recovery system and separation devices in the nose and tail.

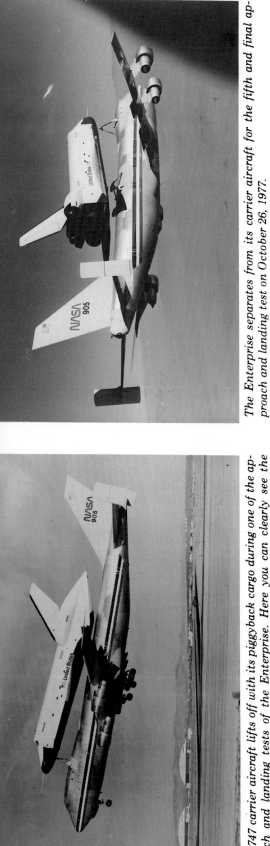

The Enterprise separates from its carrier aircraft for the fifth and final approach and landing test on October 26, 1977.

Astronauts Haise (left) and Fullerton piloted the first, third, and fifth approach and landing flights of the Enterprise. The Orbiter rests on the dry lake bed after the third free flight on September 23, 1977.

The 747 carrier aircraft lifts off with its piggyback cargo during one of the approach and landing tests of the Enterprise. Here you can clearly see the American logo still visible on the side of the aircraft.

Fred Haise and Gordon Fullerton touchdown on a concrete runway at Edwards Air Force Base during the last approach and landing test.

Two minutes after liftoff, the vehicle is already 30 miles up with with a speed of 2,700 miles per hour. At this point, the Solid Rocket Boosters burn out and separate from the External Tank. The two huge casings fall back toward the Atlantic Ocean, deploy parachutes, and are recovered for later reuse.

Just seconds after separation of the External Tank, the Orbital Maneuvering Subsystem engines are fired to inject the Orbiter into its initial elliptical orbit. On the first orbital test flight, this orbit will have a high point of 173 miles and a low point of 62 miles. About half an hour later, a second burn of these thrusters takes place to circularize the orbit at an altitude of 173 miles.

The vehicle leaves the launch pad with all engines burning. At first the huge piggyback combination of rockets, tank, and Orbiter move ever so slowly upwards. As it clears the gantry, it turns about its axis and moves perceptibly faster. Seconds later it will be accelerating, and its flight will take the bird out toward the horizon.

Eight minutes after liftoff, the Orbiter main engines are shutdown and the External Tank is separated. This occurs at an altitude of 72 miles. The huge External Tank now falls back into the atmosphere and reenters like a ballistic missile. Fragments which do not burn up fall into the Indian Ocean, almost half way around the world from the launch site.

59

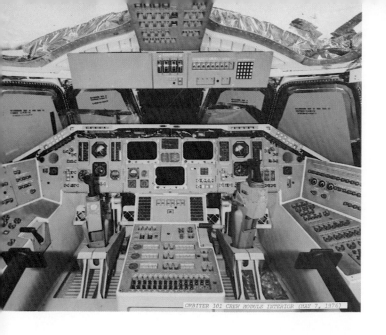

The flight deck of the Enterprise looks quite complicated, but in many ways it is similar to that of an airliner. It is organized in the usual pilot/co-pilot relationship, with duplicated controls that permit the vehicle to be piloted either from the left or the right seat. One crewmember could return the vehicle to Earth in case of an emergency.

These six Mission Specialist/Astronaut candidates are the first women to be named by NASA for the space program. They are, left to right, Margaret R. (Rhea) Seddon, Anna L. Fisher, Judith A. Resnik, Shannon W. Lucid, Sally K. Ride and Kathryn D. Sullivan. The six women, 14 other male Mission Specialist candidates and 15 Pilot candidates began a two-year training program in July 1978 and became Astronauts in 1980.

This twilight picture silhouettes the unusual arrangement of the Orbiter on the back of a 747. The complex system of cranes and other devices was used to mate the Orbiter to its carrier aircraft for transportation to the Marshall Space Flight Center in Huntsville, Alabama, early in 1978.

The following labels appear on the diagram: SEAT, WAIST RESTRAINT, ODOR/BACTERIA FILTER, URINAL, HANDHOLD, COMMODE OPERATING HANDLE, CONTROL PANEL, FOOT RESTRAINT, FAN SEPARATOR SELECTOR SWITCH, VACUUM SHUT-OFF CONTROL

The waste-collection system looks strange at first. After following a sequence of several steps, I learned how to use it. Luckily, I wasn't in a hurry at the time. On STS-3 the system got fouled up and caused a zero-g mess.

to be a combination working, living, and storage area. This pressurized compartment has a volume of about 2,500 cubic feet, and contains three levels or sections. The upper section is the *flight deck,* and it contains the displays and controls used to pilot, monitor, and control the entire vehicle and its cargo. Up to four crewmembers can be seated here. The mid-section or *mid-deck* contains provisions for three additional crewmembers or passengers, a galley, personal hygiene facilities, an air-lock, and four avionics equipment compartments. A hatch toward the aft of the Orbiter provides access to an air-lock which leads to the cargo bay. The lower section is the *equipment bay,* and it contains environmental controls for life support and provides equipment storage space. This area is accessible through hinged floor panels.

A unique waste-collection system has been developed for Shuttle crews. It has the capacity to dispose of human waste from both male and female crewmembers in the zero-G environment. The General Electric Company won the contract for this system by demonstrating a prototype model under simulated weightlessness conditions in an Air Force KC-135, which is a military version of a Boeing 707. Needless to say, this could have been a messy flight, but G.E. succeeded. Many employees and visitors at the Valley Forge Space Center in Pennsylvania have volunteered to try out the system. I was one of those fortunate visitors who found this contraption to be quite ingenious. In fact, no complaints have been registered as yet.

The nominal maximum crew size is seven. However, for a rescue mission, the mid-deck can accommodate an additional three seats with a little cramping. A total of ten seats would then be available for the three flight crewmembers and up to seven rescued personnel.

The flight deck accommodates four basic crew positions for the nominal Orbiter mission. The function of each is now briefly outlined. The *commander* and the *pilot* are in command of the flight and are responsible for the overall mission, payload operations, and safety. They must be proficient in all phases of the flight, payload manipulation, and docking. The pilot is second in command but has essentially the same duties as the commander. One *mission specialist* is included in the standard crew complement with responsibilities for the payload and associated Orbiter operations. For the more complicated missions, two mission specialists may be onboard. A *payload specialist* will not be included on all flights. However, when one or more are onboard, they are responsible for Spacelab and other scientific operations. This crewmember is not a regular NASA employee as are the others. In fact, the payload specialist may be provided by the user or some other non-government organization.

Flight deck instrument panels and controls are organized into four separate areas. There are two forward-facing, primary flight stations for the commander and pilot which make up one of these areas. Two rearward stations, one for payload handling and one for vehicle control during rendezvous and docking, represent another one of these areas. The other two control areas consist of a payload station for management and check-out of cargo and the

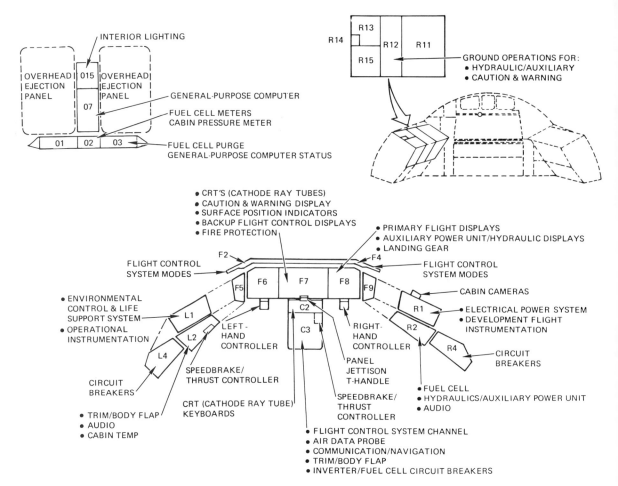

Here is a schematic of the various displays and controls on the flight deck. In the lower center, we see the various sections of the primary flight stations used by the commander and pilot. In the upper left is a schematic of the overhead segment and includes ejection panels which were available during the orbital flight tests. These panels are released in case of a true ejection situation. In the upper right is a rough schematic of the aft flight deck configuration.

mission station to monitor connections between the Orbiter and its payload and to control communications. The primary flight stations are organized in the familiar airliner configuration with a pilot station on the left and co-pilot station on the right. There is sufficient duplication of instruments and controls to permit the vehicle to be piloted from either side and to permit a one person emergency return. Dual-manual flight controls include hand-controllers which rotate to the right and left and forward and back, rudder pedals, and a speed brake actuator.

Looking to the aft-end of the flight deck, instruments and controls on the left of center (right side of the vehicle) are for operation of the Orbiter, while those on the other side are for operating and handling the payloads. This latter area contains displays and controls for manipulating, deploying, releasing and capturing satellites. Also included here are controls for opening and closing the payload bay doors and operating the large manipulator arm used for cargo handling. Two closed-circuit television monitors display video pictures for overseeing cargo bay operations. In this way, the crew stations are designed to allow a maximum of activities to be remotely controlled from inside the crew cabin. This approach tends to minimize the number of extravehicular activities required for most missions. In fact, there should be only a limited number of flights in which the crew has to venture outside of their pressurized cabin.

ALL HAS NOT BEEN ROSY

Development of the Space Transportation System began in 1972 and was originally to reach a maximum level of activity in 1976. This would have brought about an operational system in March 1979. However, schedules have been extended several times, causing a delay of over three years with a slippage in the inauguration of an operational system to November 1982. These delays are largely the result of budget limitations imposed by the

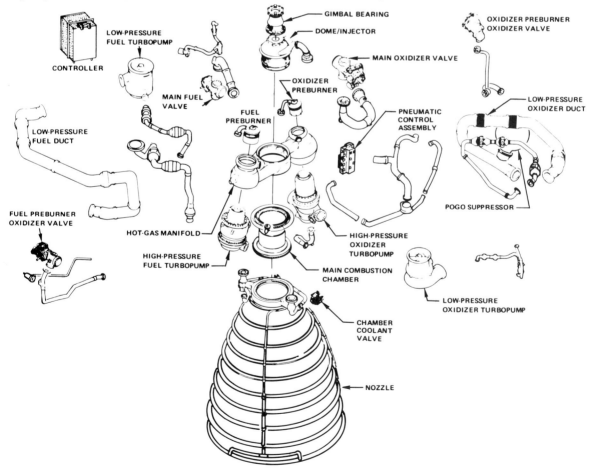

Looking to the aft-end of the flight deck we see the Mission Station on the left and the Payload Station on the right. Viewing windows to the cargo bay are in the center. Areas available for payload-related equipment are cross-hatched.

ON-ORBIT STATION
• PANEL AREA: 3.7 FT^2 (0.34 m^2)
   19 BY 13.97 IN. (0.48 BY 0.31 m)
• VOLUME: 2.4 FT^3 (0.07 m^3)

PAYLOAD STATION
• PANEL AREA: 8.3 FT^2 (0.77 m^2)
• VOLUME: 13.8 FT^3 (0.39 m^3)
   19 BY 21 BY 20 IN.
   (0.48 BY 0.53 BY 0.5 m)

CLOSED CIRCUIT TELEVISION

• PANEL AREA: 2.3 FT^2 (0.21 m^2)
• VOLUME: 1.5 FT^3 (0.04 m^3)

MISSION STATION

STOWAGE AREA

VIEW LOOKING AFT

ADDITIONAL VOLUME OF 1.3 FT^3 (0.04 m^3)
AVAILABLE FOR PAYLOAD AVIONICS BOXES

• PANEL AREA: 2.8 FT^2 (0.26 m^2)
• VOLUME: 4.6 FT^3 (0.13 m^3)
   19 BY 21 BY 20 IN. (0.48 BY 0.53 BY 0.5 m)

TOTAL VOLUME: 23.6 FT^3 (0.67 m^3)
TOTAL PANEL SURFACE AREA: 17.1 FT^2 (1.58 m^2)

Each Shuttle main engine has many components. There are two low-pressure and two high-pressure turbopumps which are key items in the successful performance of these engines.

CONTROLLER

LOW-PRESSURE FUEL TURBOPUMP

GIMBAL BEARING

DOME/INJECTOR

OXIDIZER PREBURNER OXIDIZER VALVE

MAIN OXIDIZER VALVE

MAIN FUEL VALVE

OXIDIZER PREBURNER

LOW-PRESSURE OXIDIZER DUCT

LOW-PRESSURE FUEL DUCT

FUEL PREBURNER

PNEUMATIC CONTROL ASSEMBLY

POGO SUPPRESSOR

FUEL PREBURNER OXIDIZER VALVE

HOT-GAS MANIFOLD

HIGH-PRESSURE FUEL TURBOPUMP

HIGH-PRESSURE OXIDIZER TURBOPUMP

MAIN COMBUSTION CHAMBER

LOW-PRESSURE OXIDIZER TURBOPUMP

CHAMBER COOLANT VALVE

NOZZLE

President's Office of Management and Budget. Schedule changes resulted in cost increases that reduced program reserves planned for technical and other unforeseen problems. NASA implemented a series of schedule modifications to prevent major cost overruns. Despite these efforts, a $50 million (in 1971 dollars) cost growth was announced in 1974. Further adjustments were made in 1975 and 1978 to reduce program content. Such changes allowed annual budget reductions, but they caused further delays and increased overall program costs because major tests were delayed, deleted, or reduced in scope. Even without including possible future technical problems and schedule delays, the U.S. General Accounting Office estimated in 1978 that the development program would probably exceed original estimates by $1.1 billion.

NASA has continued to work under severe cost and schedule constraints. In February 1976, it announced an additional $20 million cost growth, bringing the total announced increase to $70 million. As in previous years, NASA reduced or delayed program activities to stay within its annual budget ceiling. One result of this is that cost estimates for refurbishing the two development Orbiters and producing three additional Orbiters have increased from one billion dollars to 2.7 billion dollars. This increase has resulted in the probable elimination of one, flight-qualified Orbiter, leaving a fleet of four.

The risk of major cost growth has continued. It is possible that serious technical problems will yet be identified because some test programs have been deleted or delayed. Past experience with major civil and defense acquisitions has shown that delaying and deleting testing can lead to costly retrofit or redesign even after the system has become operational. For example, the main rocket engines on the Orbiter represent a significant advancement of technology because they are designed to be more efficient than previous engines. Furthermore, this is the first attempt to build a reusable rocket for an operational

Here is a cutaway view of the high-pressure oxidizer pump. It is similar in principle to the high-pressure fuel pump. Both of these have been redesigned at least once and have caused many delays and problems in testing of the main engines.

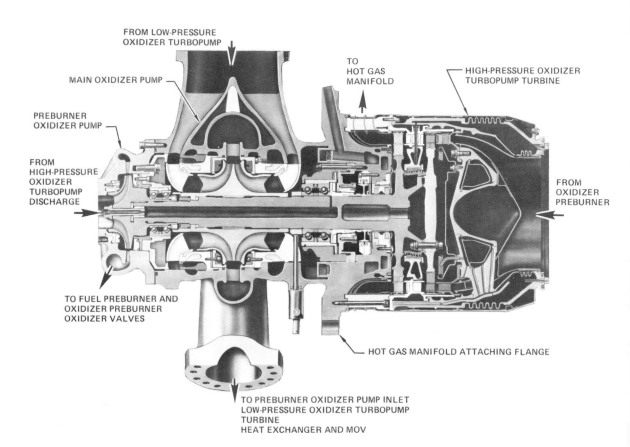

Astronauts who conducted the flight tests had an ejection and parachute system available to them. This was useable during the approach and landing tests and was available during the four orbital flight tests. Here we see tests that were conducted on an instrumented dummy at Holloman Air Force Base in New Mexico to check out the system.

space vehicle. The design goal is 55 uses before major overhaul. Serious problems in the engine development program have been found in the high pressure turbo-pumps. These are crucial to the system because they provide propellants to the main engines at extremely high pressures. For example, during critical tests, vibration caused premature wear-out of the bearings supporting the main shaft, and one of these pumps failed. This had delayed firing the engines at full-thrust to simulate actual launch operations. The problem had been under investigation for more than three years, and it could still cause problems during operational phase activities.

The solid rocket boosters are the largest ever made by the United States. Each booster, which is over twelve feet in diameter and about 149 feet long, contains over one million pounds of solid propellants. This is about twice that of previous large solid boosters used for the Titan III-C. The major uncertainty with these boosters is reusability. This is the first time solid rocket boosters have been recovered, refurbished, and reused to reduce operating costs.

SAFETY FIRST

The United States has an enviable record of safety in space since that first manned flight by Alan Shepard on February 20, 1962. We have never lost a single astronaut while in orbit. This is a reflection of the ever present concern and precautions regarding hazardous situations throughout our staffed space program. However, we can never be too safe, for on that fateful day, January 27, 1967, we did lose three astronauts in a violent fire on Pad 39 while running tests on the Apollo lunar vehicle. This will never be forgotten by the thousands of dedicated NASA personnel involved in launch operations. The sad loss of Roger Chaffee, Virgil "Gus" Grissom, and Edward White reaffirms the belief that you can never relax your safety standards or precautions. We certainly don't want to lose any more lives in the coming Shuttle years. Since April 1959, over 100 astronauts have been recruited. Of these we have lost eight, three in the Apollo fire, four in T-38 jet crashes, and one in an auto accident. In 1981 a launch site technician died during prelaunch tests of the Shuttle at pad 39A. To insure that no more launch-connected accidents occur, a comprehensive safety program has been established at each of the launch sites. This is intended not only to protect personnel and the public but also to prevent damage to property, to avoid accidental work interruptions and to help end potential future hazards. Any user of

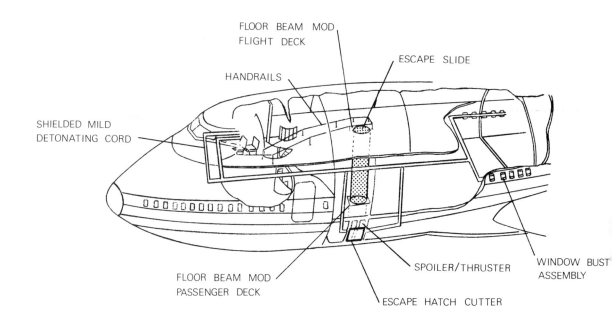

In the event the Orbiter caused damage or instability when carried atop the 747 carrier aricraft, the three crew members could escape by activating explosive circuits to open a hatch and equalize cabin pressure. They would then slide down a chute and out the bottom.

the Space Transportation System is expected and required to put into practice a long list of safety provisions regarding payloads to be launched. There is a complicated series of checks and balances to insure that these requirements are satisfied. If you are careless about handling your satellite or its hazardous contents, NASA may dismiss you from the Cape.

To explain the meaning of potentially dangerous activities associated with payload preparation, a few hazardous operations are identified here. In general a hazardous operation is one that could result in damage to property or injury to personnel because it involves one or more of the following. Any environment that deviates from normal atmosphere in terms of pressure, chemical composition, or temperature has the potential for harm. The handling, transportation, installation, or removal of explosives is an obviously dangerous situation. Propellants pose a similar threat to safety whether loading, unloading, or moving them. Any operation which involves lifting, loading, or transporting of large or heavy objects must be included in our list.

Safety precautions and procedures are extensive. Every activity during launch, reentry, and landing, at the Dryden Flight Research Center and at the Cape, have been reviewed for potential dangers, and these have been identified. The risk of accidents has to be minimized for Shuttle launch preparation. Even during the approach and landing tests, the Orbiter crews could eject at any time. The 747 crew also had a firemen's pole arrangement to allow escape out of the bottom of the aircraft. Once the crew boards the Shuttle vehicle at Pad 39A, at T minus two hours, safety procedures get even more complicated. We will next delve into the many events and contingency plans associated with a typical liftoff and ascent into orbit.

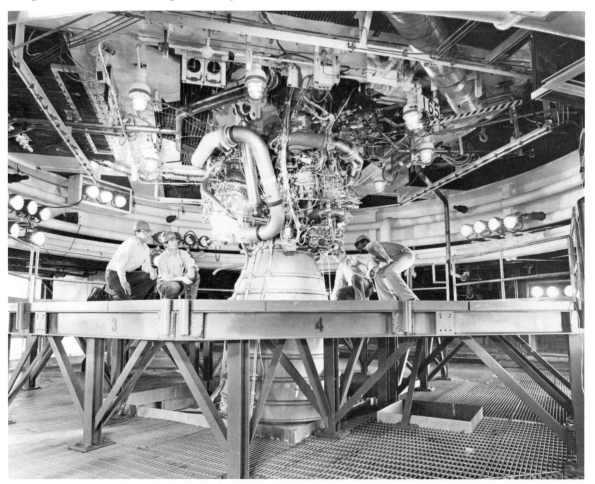

Technicians check a Space Shuttle Main Engine prior to a firing test at test facilities in Bay St. Louis, Mississippi. Each such engine is tested on stands originally constructed for the Apollo moon rocket before installation into an Orbiter.

STS-2 crewman Joe H. Engle (left) and Richard H. Truly. This picture was taken in the NASA Johnson Space Center's motionbase Shuttle mission simulator. Many hours of training are logged here in preparation for each mission.

Vice President George Bush is advised by Astronauts Young (center) and Crippen (right) as he tries out the contoured launch seat and controls of the Orbiter Columbia on Launch Pad 39A.

Just as a baby bird leaves its mother's nest, STS-1 is slowly eased out of the Vehicle Assembly Building on December 29, 1980. This began its 3½ hour journey to Complex 39, pad A. I had a once-in-a-lifetime opportunity to fly directly over the Shuttle assembly on 39A on New Year's Day, 1981.

STS-1 arrives at 39A on December 29, 1980. Almost four months later it finally leaped into space on that historic first flight.

69

Pad 39A at the Kennedy Space Center has undergone a major facelift in preparation for the Space Shuttle. Construction is now complete and the pad has been in regular use since 1981.

CHAPTER THREE

UP, UP, AND AWAY

T MINUS TWO HOURS AND COUNTING

The Shuttle vehicle is on launch pad 39A with your satellite safely stowed for its journey to the cosmos. All systems have been installed and checked. The countdown continues until it reaches T minus two hours. Here, the count is held for a short period of time to allow last minute repairs or adjustments to the various parts of the vehicle. Once the countdown passes the two hour mark, it is very difficult to stop without doing damage to some parts of the system. This last 120 minutes preceding liftoff is used to fill the External Tank with supercold liquid oxygen and liquid hydrogen. These *cryogenic* fluids cannot be stored for long periods of time, and they cause corrosion of metal parts at a highly accelerated rate. Once these have begun to fill the huge propellant tank, a hold of more than a few hours might cause enough damage to scrub the mission for weeks.

During this last phase of the countdown, the flight crew climbs aboard by ascending the launch tower and entering the vehicle through the main crew hatch. They make their way to the flight deck and take their proper positions. Of course, this is a little awkward since the Orbiter is standing on end. The crew must climb into their seats and rest on their backs while waiting for liftoff. Each one begins turning on switches and pressing buttons as they run down their respective

checklists. Ground controllers and backup teams at the Kennedy Space Center and the Johnson Space Center begin final checkout of their many TV monitors, communications equipment and computers which will assist in the liftoff and ascent into orbit. Every checklist must be completed before takeoff. Propellant crews finish filling the tanks just minutes before the engines ignite. Then all ground personnel clear the pad area. There it stands, poised to leave the Earth and voyage into space.

Ten, nine, eight, seven, six. Wait! What if something should go wrong now? From the moment the crew ascends to the entry hatch until liftoff, they are exposed to potential launch pad accidents. If, for example, the External Tank ruptures or a fire starts, the crew has two escape options. First, they will quickly release their harnesses and climb out through the hatches. One of the options is to descend on the launch tower stairs to the ground. However, if an explosion or other danger is imminent, they will use the other option. This is a *slide-wire* system for rapidly getting away from the pad. It consists of five cables extending from the launch tower to ground bunkers about 1200 feet from the base of the tower. The crew would run from the hatches to these wires. Each one supports two large fishing-type baskets, each large enough

During the last phase of a countdown, the crew climbs aboard and takes their respective positions on the flight deck. Here you see a crew of four as they would be seated in the launch configuration. The commander informs the crew that "All systems are go." Remember that they are looking up and supporting their weight on their back.

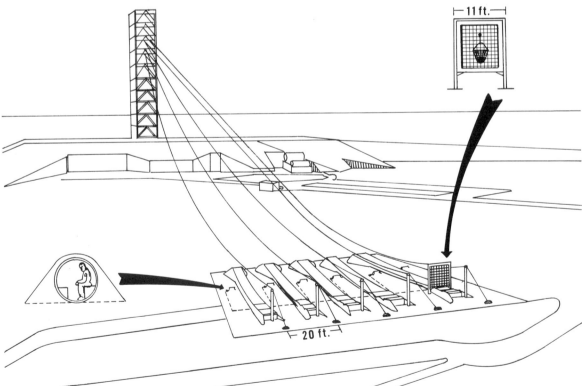

A slide-wire basket system from the launch pad to bunkers allows the crew to escape from the Shuttle vehicle in less than two minutes. Each crewmember will run from the hatches to the wires, get into large fishing-type baskets, and slide down the wire to the bunkers.

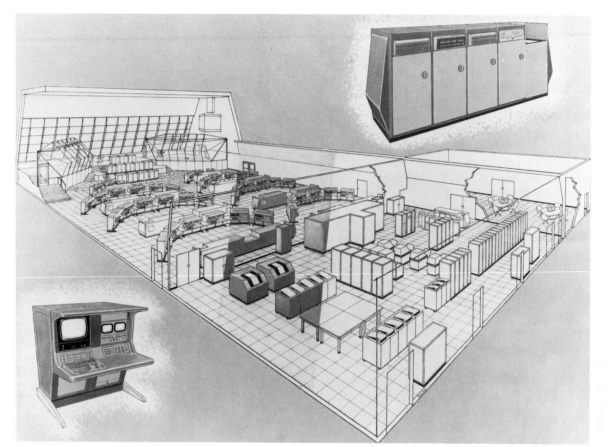

This is an artist's drawing of the firing room at the Kennedy Space Center. Space Shuttle launches are controlled and monitored from this facility. There is a large number of TV monitors and other data-handling equipment to keep continuous track of the status of all vehicle components while it leaves the launch pad and starts the ascent into orbit. Once orbit is achieved, control is switched to the Mission Control Center in Houston, Texas.

The launch team will be seated at these consoles in the firing room during the countdown and liftoff. Many functions are highly automated and team members do little more than monitor the Shuttle's performance if the launch is a nominal one.

for one crewmember. Once a basket is released, it slides rapidly down to the ground, pulled by its own weight. If it goes beyond the normal stopping point, nets will catch it. Hopefully, there will be no mishaps on the pad, and all goes well in the countdown, as it should for most flights.

Five, four, three, two, one, zero. The liftoff sequence begins with the ignition of the three main rocket engines. These should reach 90 percent of full thrust within 3.5 seconds. If they do not reach this level within 4 seconds, the main engines are automatically shut down and the launch is terminated until a later date and after a lengthy investigation of what went wrong. Let's again assume that all goes well and the main engines do reach 90 percent of their full power within 4 seconds. Then the solid rocket boosters are ignited and liftoff commences. This first 4 seconds is critical, because once the solid rockets are lit, there is no turning back. Liftoff must take place, because these huge boosters each produce almost three million pounds of thrust. If the Shuttle were not released from the hold-down clamps on the pad at this point, their force of almost six million pounds of thrust upward would probably rip the vehicle apart. Even if the main rocket engines on the Orbiter itself were to go out, the boosters would be sufficient to carry the vehicle upwards through the atmosphere. Such a contingency situation would, of course, be followed by crew ejection and loss of the vehicle. For the moment, let's concentrate on a nominal, or normal, liftoff and ascent into space.

NOMINAL ASCENT

The basic sequence of liftoff and ascent events is essentially the same for all flights. It is particularly interesting to focus on the first orbital flight which lifted off on April 12, 1981. This was known as Orbital Flight Test 1, or STS-1. It began the final phase of Shuttle development. There were a total of four such flights planned before the operational life of the system was initiated in 1982. The purpose of these flights was to demonstrate and to verify expected capabilities of the Space Transportation System under actual conditions. Let's take a look at what happened during that first orbital flight test. Its purpose was simple, to demonstrate a safe liftoff, ascent, and return of the Orbiter and crew.

Orbital Flight Test 1 had a liftoff weight of

4½ million pounds and carried a payload of only 10,000 pounds of test equipment and instruments for measuring the health of the vehicle. After liftoff, it ascended to the northeast over the Atlantic Ocean. About a minute into its flight it experienced something called

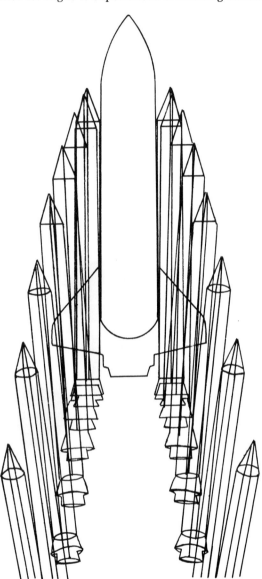

The Solid Rocket Boosters separate from the External Tank with the aid of several small rockets on the booster casings. Here we see a superimposed picture of separation dynamics with each subsequent position of the casings shown at half-second intervals.

max q, this is the point at which a maximum dynamic pressure is reached. *Dynamic pressure* is a measure of the force it would take to stop the air as it flows past the vehicle. This is a critical moment in the ascent of any launch vehicle, because it is the point at which the

For Kennedy Space Center launches, the Solid Rocket Booster casings return to the Atlantic Ocean about 150 miles downrange of the launch pad. This is well out to sea and no ships should be in danger of being hit. The search areas are relatively small for both the right and left booster casings. Actual splashdown should occur in an area 12.7x8.1 miles in size. The nozzle inserts are also jettisoned, dropping about 20 seconds after SRB shutdown.

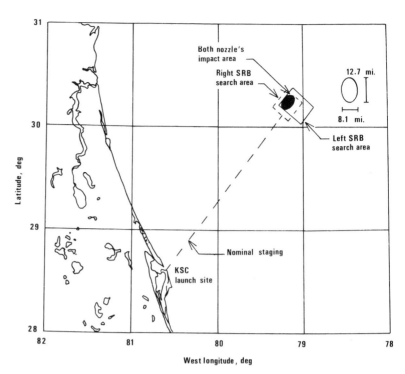

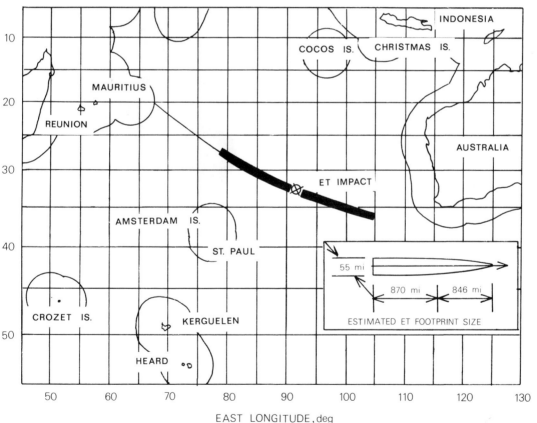

Any debris which remains from the External Tank (ET) as it reenters the atmosphere will reach an area of the Indian Ocean which is well away from any land masses. The expected area of debris impact makes a bullet shaped footprint pattern which is 55 miles wide and 1700 miles long. This represents a large area, because the reentry path is very sensitive to the tumbling motion of the tank as it leaves the Orbiter. A basic rule which is followed in targeting the tank to the splashdown area is that it cannot come within 230 miles of any land mass boundary.

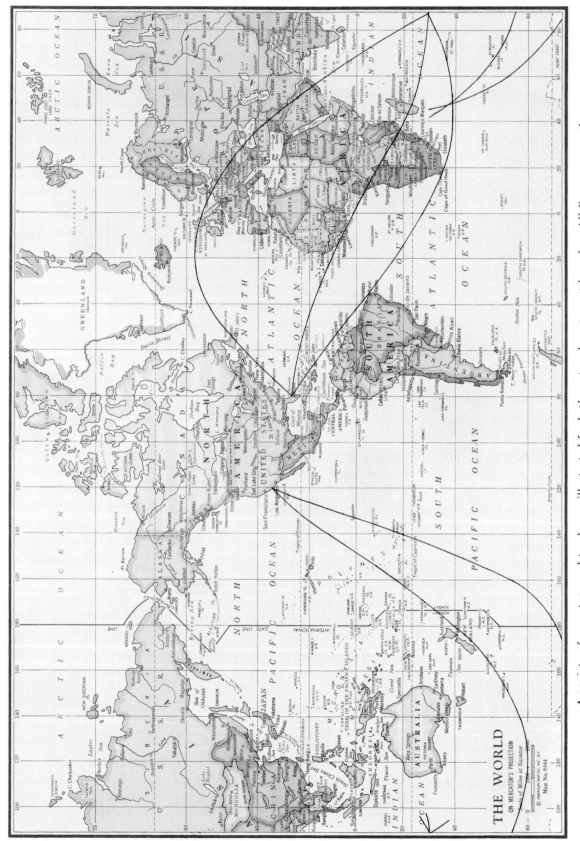

A variety of ascent ground tracks are illustrated for both east and west coast launches. All five represent the flight path of the External Tank and indicate splashdown in the Indian Ocean.

aerodynamic forces reach a peak. Beyond this point, buffeting and wind noises diminish. When this point was achieved, the vehicle was already past the 37,000-foot mark. At the T-plus-two minutes point the vehicle was already 30 miles up with a speed of 2,700 miles per hour. Here, the solid rocket boosters burned out and separated from the Orbiter and its external tank. At T-plus-eight minutes the Orbiter engines shut down and the external tank separated. This occurred at an altitude of about 70 miles. Seconds later, the Orbital Maneuvering Subsystem engines were fired to inject the Orbiter into its initial elliptical orbit. This orbit had a high point of 152 miles and a low point of 65 miles. Later burns of these thrusters circularized the orbit at an altitude of about 172 miles. This first orbital flight lasted only two days and had a crew of two men. The launch took place at 7:00 a.m. Eastern Standard Time from Pad 39A on April 12, 1981. Originally, the first launch date was to be in 1979, but delays occurred, due primarily to technical problems with the main engine development and thermal tile bonding.

Let's use modern technology for a moment to make an instant replay of the liftoff sequence. This time, let's try to imagine what the crew experienced during all of these events. The Shuttle rises slowly from the pad until the tower is cleared. Then a slight turn around its axis is noticed, and its acceleration upward is more pronounced. As the vehicle accelerates upward the two crewmen feel a downward force which increases quickly until it becomes three times their normal weight on the ground. Their arms and legs become extremely heavy. Their facial tissues are pulled back toward their ears. It is extremely difficult for them to lift their arms or legs or even to move their fingers. Nevertheless, this experience is not necessarily painful or uncomfortable. Their seats are designed to support the various parts of their bodies during this acceleration. It will last for several minutes, until the shutdown of the three main Orbiter engines.

As the vehicle lifts upward into the sky, its rockets always fire along the flight path. In order to obtain orbit, its flight path must curve so that its velocity is parallel to the horizon as orbital altitude is reached. As we watch the vehicle ascend, it rotates such that the Orbiter becomes inverted as it goes higher and higher into the sky. The crew will actually end up in an upside-down position from what we would normally expect in an airliner. However, once in space and in states of weightlessness, there is really no *up* or *down* direction; it all feels the same. At T-plus-two minutes, the solid rocket boosters burn out, and the crew feels a jolt as these huge casings separate from their vehicle. To insure that these boosters do not hit other parts of the vehicle as they separate, there are four separation rockets in the nose and four in the tail of each of these boosters which ignite on separation from the external tank and insure a safe departure for their return trip. The descent back to earth of the booster casings begins with separation of the nozzle inserts which free-fall into the Atlantic. This is followed by the opening of parachutes on the casings, bringing them slowly back to the ocean approximately 150 miles downrange from the launch pad. Two tugboats wait there, one for each of the Boosters. A plug is inserted into the end of each of these casings, and air is pumped into the empty hulk. Then they are towed back to the Cape for refurbishment and reuse.

The Orbiter continues on up into space with the sophisticated guidance system steering the vehicle to the desired main engine shutoff point. As these engines are turned off, the crew feels another jolt. The external tank then separates from the Orbiter. To insure safe separation, small rockets are fired on the Orbiter to allow it to back away from the tank, allowing the tumbling descent of the external tank to begin as it falls back into the atmosphere. It makes a reentry much like that of an intercontinental ballistic missile (ICBM). However, it does break up, and the resulting debris impacts an area in the Indian Ocean, more than 11,000 miles from the Cape.

Once the main engines are shut down, the crew begins to experience weightlessness. All that remains to achieve orbit is two small firings of the Orbital Maneuvering Subsystem. These are completed about 35 minutes after separation of the external tank.

At this point, the crew can unbuckle their harnesses and float around the cabin as we have seen them do on previous Apollo and Skylab flights. They are free to begin work on the various tests and measurements of the vehicle. This work goes on for about two days, followed by a return to earth.

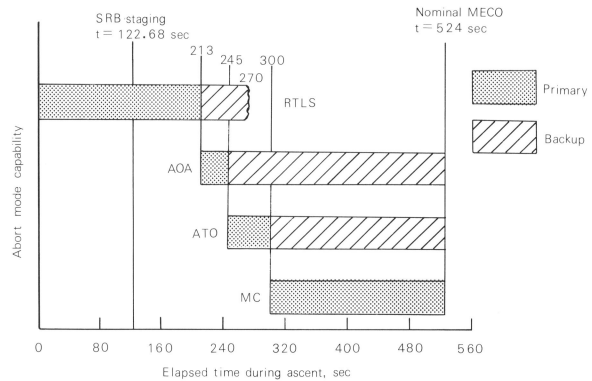

One or more of the three basic abort modes are available at any point in the ascent should a main engine fail or similar event take place. These three modes are return-to-launch-site (RTLS), abort-once-around (AOA), and abort-to-orbit (ATO). The earliest mission completion (MC) point is 300 seconds, after which a main engine could fail without requiring an abort.

ASCENT ABORTS

From the moment the crew approaches the launch pad until they climb down onto the runway after the mission, there is always a chance that an accident can occur. We have already seen that, while on the launch pad, the crew does have the capability to escape quickly up until the time of liftoff. Once lift-off occurs, the crew-escape options are somewhat more complicated. For example, the four orbital flight tests incorporated ejection seats for the crew which were similar to those used on the Lockheed SR-71 high altitude strategic reconnaissance aircraft. If a major failure had occurred during the liftoff sequence while the vehicle is still below 120,000 feet, the commander and pilot would have punched the bail-out button. This would cause ceiling panels to be blown out above their heads, followed by ejection of the two crewmen on rocket propelled seats. These devices could carry the two men safely away from the vehicle. The types of failures which might have resulted in this action by the crew included rupture of the external tank, a blowout in the side of one of the rocket boosters, complete loss of guidance, loss of thrust from one of the boosters, premature separation of a booster and so on. These ejection seats would have been used only in case of imminent loss of the vehicle. Operational flights do not have ejection seats. Thus, the crew must stay with the Orbiter from liftoff to touchdown.

There is always a risk associated with flight. A certain level of risk has always been accepted with any new venture undertaken by man. In fact, there is some risk associated with everyday activities. For example, let's take a look at commercial aircraft flights. There are always critical times during a flight in which there would be very little chance of safe escape or rescue from the airplane. If all engines on a jet aircraft fail just after takeoff, it is highly likely the plane would crash, and there would be extensive loss of life. The worst possible kind of accident is one involving the collision of two jumbo jets. Unfortunately, such an accident did occur in the Canary Islands even though the associated probability of this happening was thought to be insignificant. Because of our understanding of risk in these other situations, it is only reasonable to assume that no space program can be risk-free. To eliminate or control all potential

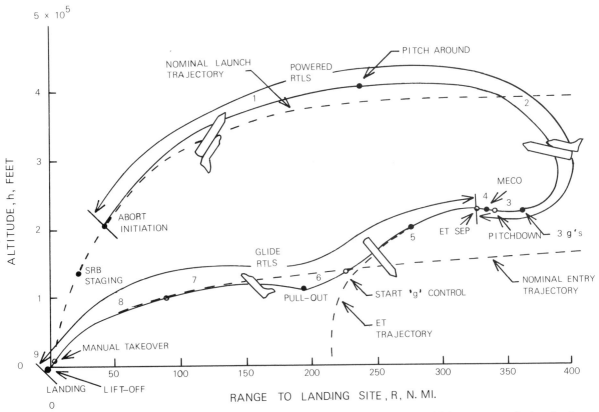

The return-to-launch-site (RTLS) abort mode can be executed for some failure which may occur during the first 4.5 minutes of flight. The sequence of events to safely return to the landing site includes a pitcharound maneuver, burnout of the main rocket engines (MECO), External Tank separation (ET SEP), and a glide to the landing runway.

hazards in the Space Shuttle Program would be a monumental task in terms of human effort and expense. A certain level of risk must be accepted to achieve an operational Space Transportation System. NASA recognizes this and has accepted a minimal level of risks in order to achieve the objectives and realize the benefits of the program.

The chances of a failure or accident occurring that required crew ejections were minimized through the use of backup systems and carefully developed procedures. Failures which do not require crew ejection can be handled in a very controlled way. Procedures have been established for the safe return of the Orbiter and crew to a landing site. Basically, three abort modes are available, if required, during each ascent sequence. These are referred to as *return-to-launch-site, abort-once-around,* and *abort-to-orbit.* They are designed to provide a continuous capability for keeping the Orbiter and the crew intact from liftoff to insertion into orbit.

The *return-to-launch-site* mode is the first one of the three available after liftoff. This strategy assures return to the runway at the Cape for an east coast launch and to Vandenberg Air Force Base for a west coast launch. It is put into operation if there is a loss of thrust on a single one of the three Orbiter main engines or for some other failure during the first 4.5 minutes of ascent which might require an immediate return. Although a return-to-launch-site abort might be required for a failure in the first two minutes, the sequence would not begin until the solid rocket boosters have separated. Thus, powered flight is continued downrange in order to loft the Orbiter and the external tank to a suitable altitude to perform a *pitcharound* maneuver. If a failure occurs after booster separation, this maneuver would be performed immediately. The pitcharound maneuver is required because the Orbiter rotates into an inverted position during ascent. Upon completion of the pitcharound, the Orbiter will be upright so that the crew will be in a *heads up* attitude, pointing back toward the launch site. The direction of flight would be reversed with sufficient speed to guarantee that the Orbiter can make it back to the landing runway. Even after booster separation, two of the three main engines will be operating to assist in the return flight. When almost all of the main rocket propellant has been used up, the Orbiter will pitch down and release the external tank. Small attitude rockets on the Orbiter will guarantee separation away from the big tank as it leaves the area. At this point the crew begins a long glide to the landing site, and the external tank falls into the Atlantic Ocean approximately 200 miles downrange from the launch site. If all goes well, the crew, Orbiter, and its payload will be safely returned.

The *abort-once-around* mode provides for a return to a selected landing site such as Holloman Air Force Base in New Mexico or Edwards Air Force Base after circling the earth once. This abort scheme can be used in case of a main rocket engine failure after 3.5 minutes into the flight, up until just prior to final orbit injection. In the case of the first orbital flight test, original considerations assumed that the landing site would be Edwards Air Force Base if such an abort mode had to be exercised. However, due to the rotation of the earth and the position of the launch site, an

additional twenty thousand pounds of main engine propellant would be required to shape the trajectory so that Edwards would be the landing site for this type of situation. Rather than waste this propellant, NASA decided to use Holloman as the primary abort landing site. Fortunately, no abort was required for STS-1. The flight profile for an abort-once-around is coincident with a nominal trajectory until the time of an engine failure. This strategy involves extra firings of the Orbital Maneuvering Subsystem rockets and the attitude thrusters before and after the main engines are shutdown.

The *abort-to-orbit* mode provides for a case such as a main engine failure after the four-minute point and up until just prior to final orbit injection. This strategy is useful for any abort situation which is not critical in terms of time. A main engine failure beyond the four-minute point may result in a low orbit, but there is no rush to get back. In general, this abort mode is intended to conserve Orbital Maneuvering Subsystem and reaction control thruster propellant for on-orbit usage. This tends to increase the possibility of at least partial mission success. In fact, if a failure occurs

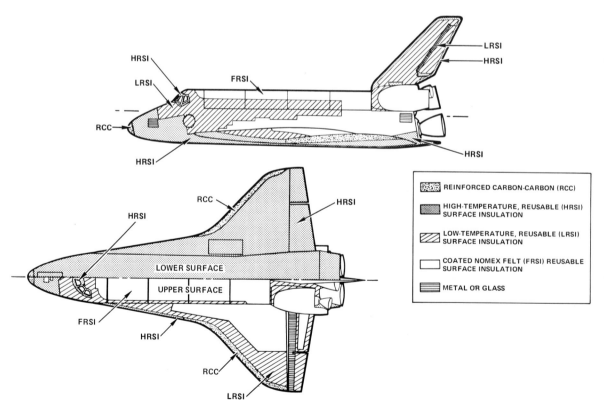

The Orbiter external skins are constructed primarily of aluminum and graphite epoxy. During reentry, these surfaces must be protected from exposure to temperatures above 350°F. Thus, the exterior is protected by a variety of reuseable surface insulation materials that must withstand up to 2300°F in certain areas.

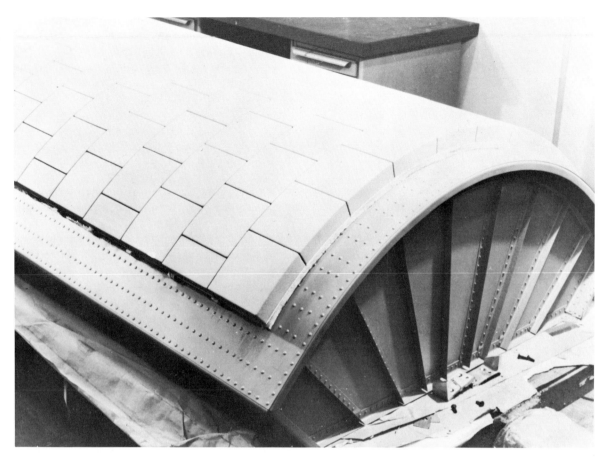

Here is a test specimen of the thermal protection system tiles or insulating bricks. These 34,000 ceramic tiles protect the Orbiter from extreme heat during ascent and reentry.

Technicians must take extreme care in gluing the protective tiles on the Orbiter wing. Only a few can be installed at one time because of the complicated process required.

just before the main engines are to be shut down, it is likely that a nominal mission can be carried out by a slight increase in usage of the Orbital Maneuvering Subsystem propellant. Of course, the ascent flight profile for the abort-to-orbit mode is again coincident with a nominal trajectory until a time of an engine failure. Once orbit is achieved in this mode, there is no urgency to return, and plans can be carefully made for the reentry and landing.

Let's try to experience the activities associated with a return-to-launch-site abort situation. Liftoff occurs at T-minus-zero. All is well and nominal through the solid rocket booster separation. About three minutes into the ascent sequence, a main rocket engine abruptly shuts down. We have just passed through an altitude of 38 miles and are about 52 miles downrange when this happens. Red lights begin flashing, and sirens are heard at the Cape, Johnson Space Center and on the flight deck of the Orbiter. Within seconds, all concerned parties know that one of the main engines has shut down. The range safety officer is alerted to the emergency situation. His job is to decide whether the population on the ground is in danger. If, in fact, the vehicle has turned around and headed for Miami instead of space, it would be his job to destruct the solid rocket boosters and external tank by remote control from the ground. Hopefully, this would follow separation of the Orbiter from the rest of the vehicle. In such situations the Orbiter would then be ditched into the ocean, and the crew would be recovered by air-sea rescue procedures. However, the case in point is one of loss of a single main rocket engine. Since this is one of the abort modes which NASA has planned for, there is no real concern for the safety of the public. The flight director on the ground, the flight commander, and the range safety officer quickly decide to carry out a return-to-launch-site abort. There is no instant change in the flight profile because so much propellant remains in the external tank that a fairly long period of burning is required for turning the vehicle around and back toward the Cape. After about two minutes of propellant dissipation and further climbing in altitude, the pitch-around maneuver is begun. At this point, the vehicle is about 275 miles downrange at an altitude of about 72 miles. The commander then pulls back on his hand controller and brings the nose of the vehicle around so that

he is soon in a heads-up orientation with the Orbiter pointed back toward the launch site. Meanwhile the rockets are still firing and turning the velocity around so that it is also in the direction of the launch site. In doing this maneuver, the vehicle will have actually reached a downrange distance of 450 miles. About 20 seconds before the main engines run out of propellant, the pilot pitches the vehicle downward into an orientation that is best for external tank separation. The main engines cut off, and seven seconds later, he pushes the separation button. The external tank falls away while the Orbiter fires small rockets to guarantee separation from the huge hulk. All thrusters are then shut down, and the Orbiter is in a gliding return to the landing site. The external tank continues to fall away and impacts the Atlantic Ocean approximately 200 miles away from the launch site. Minutes later the Orbiter touches down on the runway and is safely back at home.

UP, UP, AND OY VAY

Although the Space Shuttle was about to become a National Spaceline, there were several nagging technical and safety aspects that could have caused severe problems before the system was generally accepted as *operational*. For example, during the powered ascent from liftoff to orbital insertion, the Space Transportation System is required to have an *intact abort* capability for the selected abort modes discussed above. Failures associated with these abort modes are considered to have the highest probability of occurring. Examples include total loss of thrust from one of the main rocket engines or the loss of one of the Orbital Maneuvering Subsystem rockets. Intact abort is defined as safely returning Orbiter, crew and payload to a suitable landing site. NASA has identified other failures which are believed to have a lower probability of occurring and for which an intact abort capability has not been provided. These failures, referred to as *contingency abort* cases, would most likely require the Orbiter to ditch in the ocean. NASA personnel claim that the Orbiter should be able to "land" on water without breaking up. However, the Orbiter together with its payload would probably be damaged beyond repair. Contingency aborts include loss of thrust from more than one main rocket engine, premature Orbiter separation from the external tank, and failure of the solid rocket

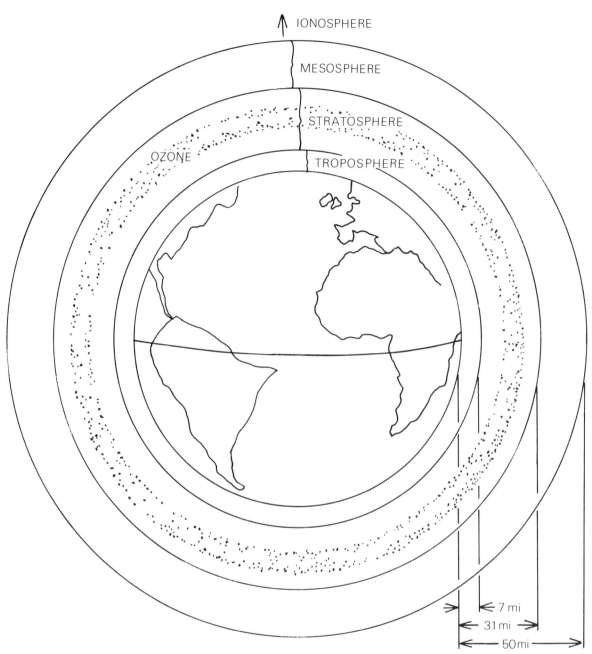

IONOSPHERE

MESOSPHERE

STRATOSPHERE

OZONE

TROPOSPHERE

7 mi

31 mi

50 mi

The atmosphere around Earth is divided into several layers for convenience of description and scientific evaluation. We live in the atmosphere which extends from the ground up to seven miles. Environmentally, the Shuttle will influence the first 31 miles of our atmosphere.

boosters to separate from the external tank. There is a third set of failures for which no provisions have been made for crew survival. NASA scientists believe that these failures, defined as *loss-of-critical-functions*, have the lowest probability of occurring because the Space Shuttle design includes redundant systems and procedures for avoiding such situations. Loss of critical functions includes major structural failure, complete loss of guidance and control systems, failure of one Solid Rocket Booster to ignite, failure of the

Orbiter to separate from the external tank, and premature solid rocket booster separation.

Let's take a quick look at what we might expect to lose in each of these situations. If an intact abort situation occurs, we can expect to waste a set of solid rocket boosters and external tank. However, the crew, the Orbiter, and the payload will probably be safely returned to some suitable landing site. If one of the contingency abort situations occurs, not only will the boosters and external tank be wasted, but both the Orbiter and payload could be damaged beyond repair. In addition, the crew could be seriously injured. In the case of loss-of-a-critical-function, NASA has not planned for any abort capability. This means that we would probably experience a total loss of the vehicle, payload, and the crew. In view of the complexity of the system and the potential for failures on early flights, NASA personnel decided to incorporate ejection seats for the two crewmembers on the four orbital flight tests.

Several problems can occur while the vehicle is still on the launch pad. For example, an electrostatic charge could build up on the surface of the external tank and possibly interfere with the Orbiter's electronic guidance and control system, providing a potential ignition source for hydrogen leaks or spills. The weather is also an important launch consideration. During inclement weather, launches will be postponed because of the possibility of lightning strikes and rain or hail impingement on the Orbiter's thermal protection layer which is the main means for preventing it from burning up on reentry. Flying through rain or hail during ascent could severely damage many of the 34,000 insulating bricks that cover a major area of the Orbiter. If this should happen, a safe reentry and return to earth might well be impossible.

Liftoff should occur about three-tenths of a second after ignition of the solid rocket boosters. During the start-up of all the rockets and liftoff, water will be pumped into flame ducts at a rate of one million gallons a minute to absorb noise and vibrations. The water suppression technique was added for several reasons. It is expected that the vibrations and acoustic noise levels experienced by the payload would exceed those of expendable vehicles without use of the water. In addition, this technique helps to absorb some of the noise generated by liftoff to the surrounding areas. All observers within a radius of several miles witnessing a typical liftoff, not only see the vehicle rise into the sky but also experience deafening noise and feel the vibrations caused by this gigantic expenditure of energy. After all, the 6½ million pounds of thrust must be absorbed by the ground and air around the launch site.

ENVIRONMENTAL IMPACT

The Space Shuttle lifts off and pushes its

way into the sky, leaving behind it exhaust, noise, and sonic booms. It is impressive, overwhelming to those who witness it in person. Most of us don't realize that a program as complex as the Space Transportation System affects our lifestyle, our technology, and our environment far beyond what we see and hear on television. Let's take a look at the *environmental impact* of our National Spaceline.

The atmosphere near the ground is affected by many Shuttle-related activities, including manufacturing, development, testing, and ascent into orbit. Higher layers of the atmosphere are influenced by launch operations since exhaust products are released all the way from the ground up to altitudes in excess of 100 miles. We will concentrate on the effects of the exhaust products on our atmosphere, however, since these are the consequences having the most rapid impact on us. Before we start, let's take a quick course in the physical properties of the atmosphere. It is largely made up of oxygen and nitrogen, although various types of other gases and particles occur at different altitudes. A vertical column of air one foot square and extending from the ground up to space weighs about one ton. About 75 percent of the total weight of our atmosphere occurs in the first seven miles. Scientists categorize air quality effects by the atmospheric layer in which they occur. These layers include the lower atmosphere or the troposphere (up to 7 miles), the stratosphere (7 to 31 miles), mesosphere (31 to 50 miles) and the ionosphere (above 50 miles). Environmental effects in the lower atmosphere are concerned primarily with chemically active substances released into the air. In the stratosphere, the major concern is with the ozone layer. Other minor effects occur at higher altitudes, but we are concerned primarily with effects up to about 31 miles.

The Space Shuttle System is powered by five chemical rockets which produce thrust by the combustion of a fuel and an oxidizer. As the vehicle lifts off from the pad and ascends through the atmosphere, it drops a total of 3.7 million pounds of exhaust products into the air. The main environmental pollutant near the launch site results from the combustion of the solid rocket boosters. By the time the Shuttle has reached an altitude of 30 miles, it has dropped most of its combustion products. As the vehicle accelerates to higher and higher speeds through the atmosphere, the exhaust products tend to be deposited in a thin column rather than being spread out as they are near the launch pad. This column diffuses into the air. At the lower altitudes, a cloud of exhaust products is generated. This is called a *ground cloud,* and it disperses slowly, making it the subject of extensive investigation. In a normal launch, this ground cloud is formed at the base of the launch platform. It consists of hot exhaust products from the solid rocket boosters, the main propulsion engines, steam from launch pad cooling, and some sand and dust drawn into the cloud, from the surrounding area. Because of the high temperatures in the cloud, buoyancy effects causes it to rise to altitudes of up to two miles where it stabilizes as the gases cool. Several chemicals are held in this cloud which are potentially dangerous to animal and plant life. For example, a typical cloud would contain over 100,000 pounds of aluminum oxide, 77,000 pounds of hydrogen chloride, 5,000 pounds of nitrogen oxides and 8,000 pounds of chlorine. On the positive side, it contains about a quarter million pounds of water vapor. Nevertheless, I suggest that if you see one of these clouds coming your way, get out of its way.

Three areas of environmental concern are associated with the ground cloud: toxic substances from the cloud, acidic rain, and inadvertent weather modification. Toxic substances of primary concern are hydrogen chloride, chlorine, nitric oxide, nitrogen dioxide, and aluminum oxide. Fortunately, the surface concentration of these substances for the Shuttle ground cloud are expected to be less than established human-exposure limits. In other words, if you happen to be standing under one of these clouds as it passes by, there should be no adverse effects. Of course, if I had a choice, I think I would move. Some sun sensitive plant species may experience discoloration due to the exposure of the diluted hydrogen chloride. Conditions inside the cloud could be drastically different than those on the ground. For example, at altitudes of several thousand feet, there may be concentrations of hydrogen chloride in excess of toxic levels. So, if you happen to be flying a small plane toward one of these clouds, I suggest you definitely divert your flight path.

Acidic rain is a widespread, low-level form of pollution, occurring as a result of burning fossil fuels. The mechanisms by which rains become progressively more acidic are not totally understood. The Shuttle cloud can lead to a special type of acidic rain caused by the solution of hydrogen chloride into the water droplets. Depending on atmospheric conditions, the exhaust cloud may entrain enough water to generate a light rain or mist. The cloud could encounter rain from a higher layer of clouds. The rain or mist precipitated as a result of this would be acidic. Two incidences have been reported which involve this phenomenon due to the firing of solid rocket motors. The first incident occurred on June 17, 1967, when a large, solid-propellant rocket motor was fired in Dade County, near Homestead, Florida. The test was made during shower activity accompanied by gusty winds. Acidic rain fell on lime and avocado groves six miles from the test firing site. The crop experienced some damage, the fruit was spotted and considered to be unsalable. The second incident occurred at the Kennedy Space Center in September 1975 during the launch of a Titan booster. Thunderstorms moved over the exhaust cloud just minutes after liftoff. A NASA environmental monitoring team was on the site measuring acidity values. The results indicated that the resulting rain had an acidity level corresponding to those of normal human stomach fluids. I'm not really sure whether this is good or bad, but I do know that our stomachs are pretty good at digesting a variety of vegetables and meats. The degree of acidity of rain generated by the Shuttle is a subject of ongoing research at NASA centers. Results so far indicate that it would be highly localized and temporary.

Inadvertent weather modification by Space Shuttle exhaust is a difficult thing to assess. This is not surprising if you consider the limited success with which our meteorologists predict local weather for one or two days in advance. Nevertheless, research in this area suggests that individual Shuttle ground clouds might modify local weather for up to two days after liftoff. The area that might be affected is estimated to be confined to an area of less than eight miles in radius. The nature of any modifications might include the intensification or the suppression of rainfall. No long term or large scale weather changes should be expected.

As the Shuttle ascends into the stratosphere, chemicals are introduced into the atmosphere which could reduce the mean level of ozone. Because of the long residence time of gasses in the stratosphere (several years), resulting effects would not be confined just to the launch site area but would also be distributed over the globe. The stratosphere extends from 7 to 31 miles up and is the region in which the supersonic transports fly. Several years ago the hysteria over the depletion of ozone by the U.S. Supersonic Transport was a major contributing factor in its cancellation. It is now generally believed that these effects were highly exaggerated. It is important to take a careful look at the situation with respect to the Shuttle. In contrast to the lower atmosphere, where turbulence and vertical mixing occur, the stratosphere is relatively quiescent. As a consequence, it is particularly susceptible to contamination because pollutants introduced there tend to remain for long periods of time. One of the *trace* constituents of the stratosphere is ozone. Although this gas represents only a few parts per million of the substance which makes up the stratosphere, potential threats to its concentration have become a focus of scientific interest and public concern for several years now. Even in its small amount, stratosphere ozone absorbs most of the biologically harmful solar ultraviolet radiation. This prevents the harmful rays from reaching the ground in quantities which could adversely affect human beings, plants, and animals. Ozone also absorbs infrared rays, and this plays an important part in maintaining the heat balance of the globe. Extensive measurements of the total amount of ozone present in the atmosphere have been carried out during the past four decades. Results vary considerably with positions over the Earth and with the time of day. These changes are relatively large and regular in character, and their causes are generally well understood.

As mentioned above, concern over altering the average level of stratospheric ozone was raised as a possible consequence of the emissions from aircraft flying at high altitudes. Effects of aircraft and other activities such as Shuttle launches on the ozone concentration depends on the natural processes that determine the distribution of ozone in the stratosphere. Fortunately, these natural processes are well understood. The ozone distribution is maintained as a result of a continuing dynamic balance between creative and destructive forces. For example, ozone is produced in the upper stratosphere by the action of solar ultraviolet radiation upon oxygen

molecules. It is destroyed by several processes. One of these is a chain reaction involving various oxides of nitrogen. Other relevant destruction mechanisms include the reaction of oxygen atoms with ozone and chain reactions involving hydrogen, chlorine, and nitrogen. In other words, the Shuttle cannot enhance production of ozone in the stratosphere, but it can enhance the destruction of this valuable gas. Thus, Shuttle launch operations will result in a net decrease of the ozone in the stratosphere. At a maximum launch rate of 60 per year, the Space Transportation System is predicted to cause a reduction in ozone concentration over the northern hemisphere of only about one-half of one percent. Over the southern hemisphere, this effect is predicted to be less than one-tenth of one percent. Nearly all of this reduction is due to the presence of chlorine compounds in the exhaust.

The biological significance of the estimated ozone reduction due to full-scale operation of the Space Shuttle is that there will be an increase in the level of biologically harmful ultraviolet radiation that reaches the ground. For small changes in ozone concentration, the percentage increase in this radiation reaching the surface is approximately twice the ozone reduction. In other words, a one-half percent reduction in ozone will result in a one percent increase in this radiation at the ground level. The biological effects of this increase can be divided into those concerning plants and animals and those concerning human beings. A one percent increase in ultraviolet radiation will have no measurable effects on plants and animals. The most significant potential effect on human beings is an increased incidence of skin cancer. While it is entirely possible that the ozone reduction and the associated increase in radiation due to Shuttle operations may lead to some increase in the incidence of skin cancer among susceptible individuals, such an increase will not be detectable because of the great variability of this radiation over the earth's surface. In other words, any increase in the incidence of skin cancer would more likely be due to the natural variations of ultraviolet radiation.

Let's concentrate for a moment on another important environmental effect. *Acoustic noise*, which is differentiated from sonic booms, is generated in many different aspects of the Space Shuttle Program. The major noise effects are generated by rocket engine

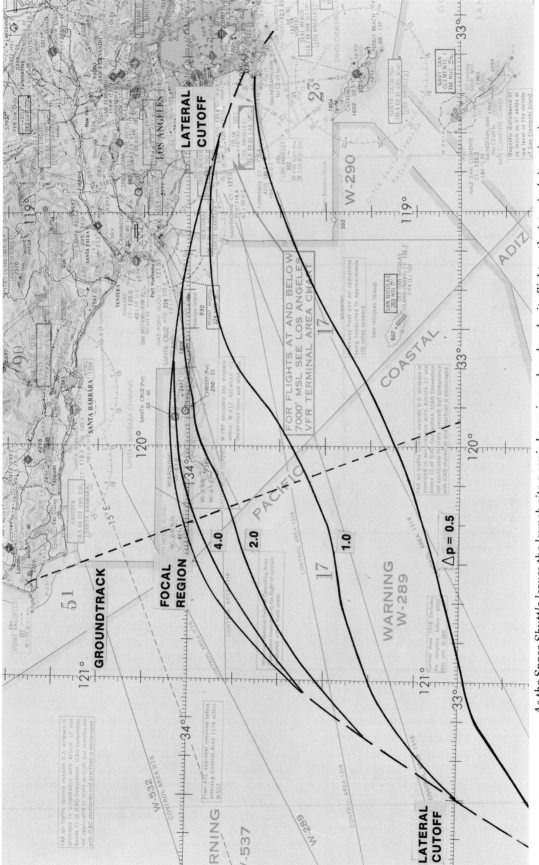

As the Space Shuttle leaves the launch site a sonic boom is produced under its flight path. A typical situation is depicted for Vandenberg Air Force Base. Maximum overpressures occur at the Channel Islands which are southwest of Santa Barbara, California. Launches to the southeast may be restricted because of potential sonic boom impingement on the Los Angeles area. Overpressures are given in pounds per square foot.

Sonic boom overpressures (pounds per square foot)	Behavioral effects
0.0 to 0.3	Orienting, but no startle response Eyeblink response in 10% of subjects No arm/hand movement
0.6 to 2.3	Mixed pattern of orienting and startle response Eyeblink in about half of subjects Arm/hand movements in about a quarter of subjects; no gross bodily movements
2.7 to 6.5	Predominant pattern of startle responses Eyeblink response in 90% of subjects Arm/hand movements in more than half of subjects; gross body flexion in about a fourth of subjects
7.1 to 13.4	Arm/hand movements in more than 90% of subjects

testing and launches. The liftoff thrust of the Space Transportation System is almost 7 million pounds which is slightly less than that of the Apollo-Saturn V thrust of 7.5 million pounds. The Shuttle produces nearly the same sound levels as did the Saturn V during a typical liftoff. During the first minute of liftoff and ascent, sound levels will be intense. In fact, observers seated in Kennedy Space Center viewing stand about 3.7 miles from the pad should have ear protection devices or at least should hold their hands tightly cupped over their ears during liftoff. The expected sound level will be above the *threshold of pain*. The viewing stand itself will vibrate. Many of the observers may find it difficult to breathe because of the tremendous energy being released at a rhythmic rate which may interfere with normal breathing processes. Fortunately, this exposure will only last for up to one minute. The nearest mainland area from Launch Pad 39A is 11 miles away. At this point, the sound pressure level is greatly reduced and is more bearable during a typical Shuttle liftoff. To summarize acoustic noise effects during Shuttle liftoffs, sound levels in regions accessible to the public are below limits suggested by the Environmental Protection Agency of the United States. Low frequency sound may cause minor damage to privately owned structures outside the immediate launch site areas, but experience from previous operations indicates that damage claims will be small and very infrequent.

As a "body" moves through the air, the air must part to make way for that body. After its passage, the air must again occupy the space through which the body passed. In flight up to speeds of about 700 miles per hour, pressure signals which are waves that travel at the speed of sound move ahead of the body to forewarn of its approach and to assist in the parting of the air. Thus, the passage of a body is a smooth process. In supersonic flight, the vehicle travels at a speed faster than the speed of sound. In other words, it travels faster than the pressure waves can forewarn the air ahead. So the parting process of the air is abrupt. In fact, this is what causes the phenomenon of *shock waves*. These waves travel through the atmosphere as pressure waves and, because of the abrupt noise they generate when passing observers on the ground, are called *sonic booms*. As the Shuttle vehicle ascends into the atmosphere, it quickly exceeds the speed of sound, generating a sonic boom as it ascends into orbit. A similar phenomenon occurs with the reentry and approach to landing. A typical sonic boom has some of the characteristics of an explosion. Even at great distances from the vehicle, where pressure levels produced are physically harmless, some public complaints are received. To understand the many factors that influence the strength of the sonic boom, we need to consider what happens as the Shuttle leaves the launch pad and accelerates to supersonic speeds.

Of course, as the Shuttle leaves the launch pad, and for the first minute or so, no sonic boom is produced because the vehicle has not exceeded the speed of sound. Maneuvers associated with the flight path result in a focusing of the shock waves over a small area of the surface just below the flight path. Intensity of the boom can be greatly exaggerated in this *focus* area. In fact, the allowable flight-path directions away from the Shuttle launch sites

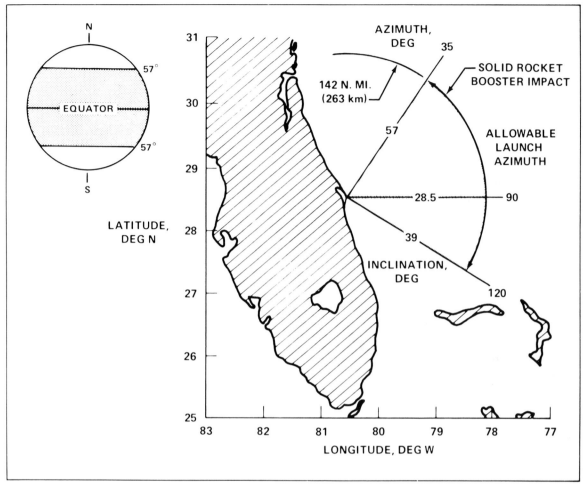

Most Shuttle flights will be launched from the Kennedy Space Center. Orbital inclinations reachable from there vary between 28.5° and 57°. The inset globe illustrates the extent of coverage possible over the Earth. The launch azimuth is the ascent direction measured from true north. An azimuth of 35° results in an orbit with inclination equal to 57°, and so on.

are restricted to unpopulated areas largely due to the intensity of the associated sonic boom. The maximum intensity is experienced at about 45 miles downrange of the pad. Beyond this point the Shuttle is climbing higher and higher, and the strength of the pressure pulses diminishes.

The sonic boom is an impulse noise, defined as a discrete noise of short duration, in which the sound pressure level rises very rapidly to a peak. Explosions have many of the same characteristics. A natural phenomenon which bears a striking resemblance to sonic booms is the thunder produced by lightning strikes. Many times the two are indistinguishable, the difference being that we expect thunder claps, but sonic booms tend to be unexpected. Impulsive noises which are novel or unexpectedly loud can startle people and animals. Even very mild impulsive noises can awaken

sleepers. Because startling and alerting responses depend largely upon individual circumstances and psychological factors unrelated to the intensity of the sound, it is difficult to make any generalization about acceptable levels in this connection. Typically, sonic boom strength is measured in terms of something called *overpressure*. For example, the overpressures inside a car when the door is closed can be as high as 4 pounds per square foot for standard sedans and station wagons and up to 8.5 pounds per square foot for compact cars. Overpressures of 12 pounds per square foot have been measured in public viewing areas during fireworks displays. The National Academy of sciences has established criteria for impulse noise limits. These are levels above which inner ear damage and hearing loss can be experienced. The limits established for one impulse per day corresponds to an overpressure level of 7.6 pounds per square foot. Thus, if you have a compact car, you should always open the window a little before slamming the door.

Public acceptance of sonic booms below the impulse noise limit is very complex and involves not only the physical stimulus but also various characteristics of the environment along with any experiences, attitudes, and opinions of the population exposed. Extensive research was done on this phenomenon during the 1960s. Results indicated that the effects of a single sonic boom on human beings can be grouped into four broad classes according to the range of overpressure.

NASA scientists designed the ascent trajectory so that any overpressure experienced in populated areas will fall in the first classification. In other words, any sonic booms experienced along the Florida coast will be quite similar to a mild thunder clap. In the case of launches from the west coast, residents near the coast in the Los Angeles area may experience the same kind of thing. To insure that no people are exposed to the focused part of the sonic boom pattern during ascent, the range safety officer at each launch site designates a *launch danger zone* for each lift-off. This comprises a sea area and an air space measured from the launch point and extending downrange along the intended flight path. The size is based on the potential hazard to ships and aircraft. Helicopter and radar surveillance of this zone commences an hour before launch. Should the overpressure levels be considered harmful, the location of the focus boom will be included in the launch danger zone. Ships and aircraft in the area likely to be affected will be warned of the pending launches. This is also the current practice for expendable launch vehicles. For some launches out of the Vandenberg Air Force Base, a focused sonic boom will strike the Channel Islands. This is a group of small islands to the southwest of Santa Barbara, California. The islands are well-known as a breeding ground and habitat for many rare

During the first launch (STS-1) an unexpected pressure wave caused minor damage to the Orbiter. This wave occurred when the SRB's were ignited. To avert further damage, NASA devised a water spray concept to abort this pressure pulse. It works like a charm.

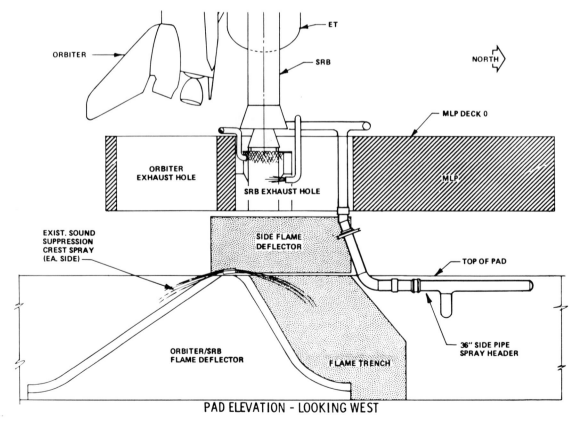

PAD ELEVATION - LOOKING WEST

Columbia rises off the pad a few seconds past 7 AM on April 12, 1981, bringing the dawn of a new age of spaceflight. It gave me goosebumps all over and I could hear the Star Spangled Banner in the background.

After the three Orbiter main engines shut down, the External Tank is jettisoned. To insure safe separation, small rockets are fired on the Orbiter to allow it to back away from the tank. The huge empty shell soon begins to tumble as it falls back into the atmosphere. It makes a reentry much like that of an intercontinental ballistic missile. Most of it burns up, but some debris does reach the Indian Ocean.

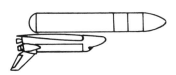
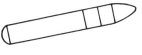
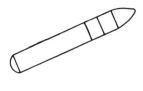
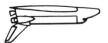

(a) RELEASE	*(b) t = 14 sec*	*(c) t = 28 sec*

These are the nozzle plugs used to recover the SRB casings from the Atlantic Ocean after splashdown. A 600-foot umbilical cord with electrical wires and air hose connect each nozzle plug with a retrieval tug. Underwater TV cameras, mounted on the nose of each plug, permit a remote operator to steer the device directly into the booster's nozzle. Metal arms extend to lock the plug in a watertight position. Water is pumped out and the booster is towed back to the Cape.

and endangered species, including the Brown Pelican. Overpressure levels could reach 30 pounds per square foot on these islands. Recently, marine scientists and environmentalists confronted Air Force officials with their concerns about possible danger to these species as a result of intense sonic boom exposure. They claimed that adult birds and marine mammals could be forced to abandon their breeding areas, leaving the young prey to predators or starvation. In the case of seals and sea lions, whole populations could be so startled that the pups might be trampled as the adults stampede toward the sea. The Air Force has conceded that this is a difficult situation because reduction in sonic boom intensity would mean severe reduction of the Space Shuttle capability for launches out of the western United States. On the other hand, it is almost impossible to devise a protective scheme for the animals on the islands without disrupting their way of life. It will certainly be interesting to see how the situation is resolved.

ORBIT ACHIEVED

If all goes well; if the Orbiter main engines perform properly; if the solid rocket boosters separate at the right time; if the external tank separates at the proper moment; and if the Orbital Maneuvering Subsystem fires its first burn correctly, then orbit is achieved about 11 minutes after liftoff. Another half hour of coasting goes by, then the final small burn inserts the Orbiter into its parking orbit. At last, the Shuttle is free of gravity, free of the atmosphere, traveling at speeds of over 17,000 miles per hour and not feeling it. The sensation of orbital flight is indeed unique.

Next we will experience a variety of orbital operations, including extravehicular activities and the handling of payloads. We will learn the number and kinds of things that can be done in space with the Shuttle. After all, this is the payoff for the many years and billions of dollars spent in developing the National Spaceline.

The two SRBs used on STS-1 are towed home by the tug UTC Freedom. A second tug, the UTC Liberty, is also used in the SRB recovery operation.

One of the Solid Rocket Boosters used on STS-1 is "dewatered" at sea. When lowered properly by parachute to the Atlantic, these expended casings fill partially with water and float in a vertical mode. Special nozzle plugs are inserted in the rocket throats and this water is pumped out. As the water leaves, the casing becomes horizontal and is towed in by tug to Cape Canaveral Air Force Station.

Once orbit is reached, the crew moves to their respective stations on the flight deck. The payload bay doors are opened, and satellites are prepared for release via the manipulator arm.

CHAPTER FOUR

OPERATIONS IN SPACE

VACUUM OF THE COSMOS

Both of the Orbital Maneuvering Subsystem rockets are shut down once the Orbiter has established its circular orbit, typically about 180 miles above mother earth. The crew sits speechless for a moment to reflect this unique sensation. Imagine for a moment how they feel and what they see. Although strapped to their seats, they feel the relaxation of weightlessness. The co-pilot cannot resist the temptation to let go of his pencil to test what he knows already. Yes, it gently hovers there, ever so slowly drifting away.

Visually, it is difficult to determine whether you are in space or somewhere else. The instrument panel shows a variety of indicator lights, computer readouts, and dial settings. There is no big neon sign saying, "You are now in space." It is not until you turn down the inside lights and look out the viewing ports that you realize where you are. Looking away from earth, you see black nothingness, except for the uncountable light specks which make up the Universe. Looking at earth, you may see a dark sphere with widely scattered patches of light, the cities at night. If it is daytime, there is the most beautiful blue, green, and brown planet with scattered white cloud areas. It is almost unbelievable. The smallest identifiable thing is a super-highway or a large river. Don't bother trying to find your house; you'd be lucky to find your town.

Sounds are deceptive, for in a vacuum noise cannot be carried, transmitted, or even created. Within the Orbiter, you hear an occasional muffled clicking due to the reaction thruster valves opening and closing in order to maintain Orbiter orientation. The oxygen circulation system and other life support components can also be heard. Otherwise, all is quiet.

THE JOB IN SPACE

The Orbital Maneuvering Subsystem engines shutdown after the second burn to finalize the parking orbit. A *circular orbit* is just that, the Shuttle maintains a constant altitude above the ground. Each orbit also has properties characterizing its orientation in space. Two quantities, *inclination* and *celestial longitude*, give its orientation relative to the stars. Inclination is the angular deviation of the orbit from the earth's equator. Thus, an equatorial orbit has an inclination of zero degrees, while a polar orbit has one of 90 degrees, and so on. The celestial longitude gives the location of the equatorial crossing of the orbit. The inclination is a result of the launch site location and the ascent flight direction as the launch vehicle leaves the pad. Celestial longitude is a result of the launch time. Of course, such orbital properties are carefully selected and determined by the mission objectives and the payload design.

Shortly after the orbit is established on a

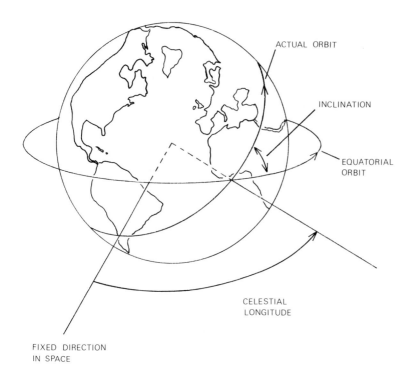

ACTUAL ORBIT

INCLINATION

EQUATORIAL ORBIT

CELESTIAL LONGITUDE

FIXED DIRECTION IN SPACE

Inclination and Celestial Longitude tell us the orientation of an orbit with respect to the stars.

typical flight, the cargo bay doors are opened. This is followed by a long sequence of checkout and turn-on steps if a payload is to be released. The Orbiter may even orient itself so that the bay is facing earth to allow direct transmission between the payload and earth as ground stations are passed over. Actual release may be accomplished by the *Remote Manipulator System* or a set of ejection springs which provide a separation speed for rapid departure. The Remote Manipulator System is a long skinny arm with joints at various points which can be manipulated from inside the Orbiter crew cabin. Each Orbiter is fitted with one of these arms to be used in manipulating payloads within the cargo bay. Such devices have been used on a smaller scale for the handling of nuclear materials on earth. However, this is the largest one of its kind. If the payload is to be ejected by springs, the mission planners will have certainly requested that the Orbiter be turned to an attitude before release which will send the payload in the desired direction upon separation. If, for example, the satellite has an upper stage or kick motor attached to achieve a higher orbit, then its ejection orientation is important to the successful ascent to a final orbit.

What is the crew doing during these events? As soon as a good orbit is confirmed by ground controllers, a series of transitional activities and responsibilities takes place. It is a time of change, from the ascent phase to the mission

phase. The commander hands over authority to the mission specialist who is responsible, along with the payload specialist if on board, for all payload and cargo bay activities. These specialists have a variety of controls at their fingertips. Not only can they manipulate cargo, but they can also orient the Orbiter as needed and turn on television cameras, lights, and payload systems. Each satellite to be launched must checkout okay before release; otherwise, they must fix it or bring it back. The checkout process is long and complicated for most satellites. It may take hours of tests and data processing to complete. However, this is where the Shuttle pays off. When the satellite leaves, we are assured of a successful mission.

The Space Shuttle has a unique capability to respond to national and worldwide needs with the flexibility to meet the requirements of changing technology amd new innovations. As we know, its primary mission is to deliver payloads to earth orbit. According to its design intent, the Shuttle will not be limited to uses originally foreseen but is ready for expanding activities in space during the 1980s. Satellites launched on previous missions may be retrieved and returned to earth for refurbishment and reuse by the Shuttle. As many as five individual satellites can be delivered on a single flight. Most retrieval flights will involve only one satellite capture. Some of these could be non-U.S. spacecraft. For example, a Russian satellite is launched and quickly

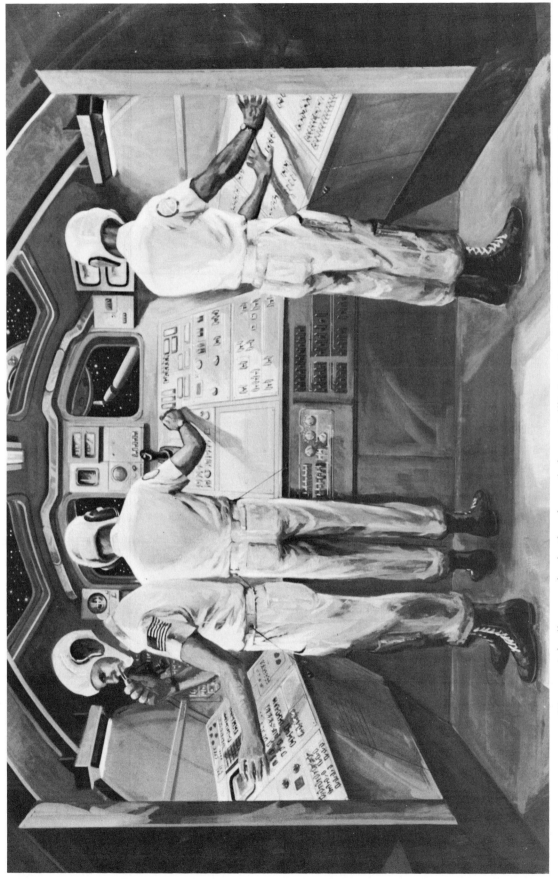

The crew is busily preparing for the release of a large satellite from the cargo bay. The mission specialist carefully maneuvers this payload with the remote manipulator.

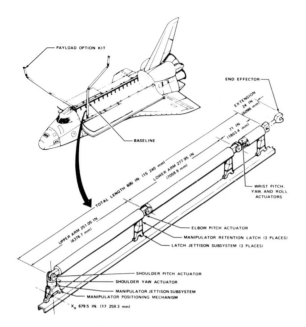

The Remote Manipulator System is a 50-foot-long arm with joints at various points to permit its use in moving payloads into and out of the cargo bay.

malfunctions. It is possible that a Shuttle flight could retrieve and return it for repairs, saving the Soviets quite a few rubles. To recover a satellite, the Orbiter will rendezvous with it, maneuver to a position close by, and either seize it with the remote manipulator arm or send an astronaut or a robot-like craft to capture it for return to the cargo bay.

The Large Space Telescope will be launched by the Shuttle and become a free-flying instrument for use by astronomers around the world. Controlled from the ground, it will open a new window on the universe. Present earthbound telescopes are severely restricted in their penetration of deep space due to the limiting properties of our atmosphere. Once in orbit, the only limitations are those imposed on the telescope itself. We can expect to see ten times as much with this new space telescope. Upon delivery to orbit, it will be prepared by mission and payload specialists for operation after release. There will be scheduled revisits to this deep-space facility, during which crewmembers will service it and exchange scientific hardware. Several years later, a Shuttle flight will return it to earth for refurbishment or retirement, possibly to the Smithsonian Museum.

Among the many payloads carried up by the Shuttle is a device which can be likened to a free-floating buoy that is carried by the ocean currents and is used to record the natural activities of weather and oceans. In the orbital case, the Long Duration Exposure Facility will be released from the Orbiter to float freely in space above the earth. It is a reusable platform on which research-type instruments will be mounted. Since it is simply a structural framework and passive in its motion, it is a low-cost way to fly a variety of experiments which collect data on the effects of the space environment over a relatively long period of time. After an extended period in orbit, it is to be retrieved and returned for experiment analysis.

Most free-flying payloads will be actively controlled, either from the ground or automatically. These will be journeying to a variety of earth orbits and deep space targets. Travel beyond the Shuttle orbit will require propulsion stages for boosting the payload to higher altitudes or to escape the earth's pull altogether. The payload will consist of a satellite/propulsive stage combination, delivered to orbit and deployed by the Shuttle. After release, the Orbiter moves a safe distance away, followed by a radio signal to fire the rocket engine. With this, the payload quickly departs from the vicinity of the Orbiter and is on its way. The view from the Shuttle should be quite interesting. In a matter of seconds after rocket start-up, the payload would accelerate out of sight. The crew will be impressed even though no sound will be heard. I imagine it will be like viewing a drag race from the starting line on television with the sound turned off.

Introduction of the Space Transportation System has permitted new thinking for those in the spacecraft building business. NASA

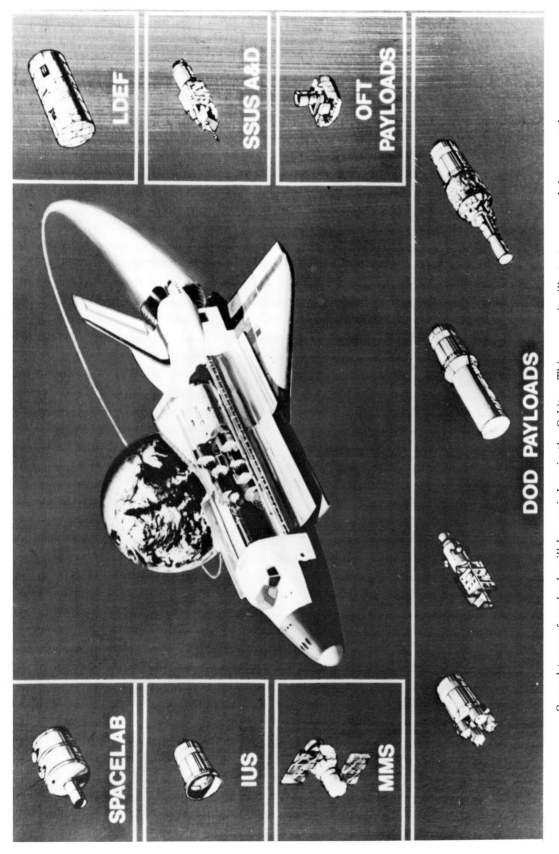

Several types of payloads will be carried up in the Orbiter. This composite illustrates some of the expected repeat payloads. These include Spacelab, Inertial Upper Stages, Multimission Modular Spacecraft, Department of Defense payloads, Long Duration Exposure Facility, spinning solid upper stages, and orbital flight test (OFT) payloads.

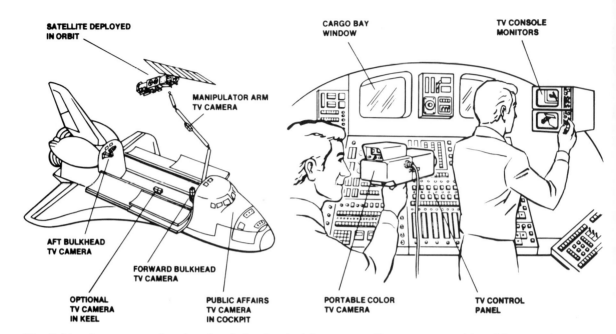

SATELLITE DEPLOYED IN ORBIT

MANIPULATOR ARM TV CAMERA

AFT BULKHEAD TV CAMERA

FORWARD BULKHEAD TV CAMERA

OPTIONAL TV CAMERA IN KEEL

PUBLIC AFFAIRS TV CAMERA IN COCKPIT

CARGO BAY WINDOW

TV CONSOLE MONITORS

PORTABLE COLOR TV CAMERA

TV CONTROL PANEL

The Orbiter has a comprehensive closed-circuit television system. Cameras are positioned in many locations around the vehicle. Television controls are located at the Mission Specialist's panel.

personnel have recognized this and have initiated the development of a multi-purpose, reusable satellite design. The concept utilizes modular components in a family of combinations of hardware pieces. Resulting from this project is the *Multimission Modular Spacecraft* which will be ringing in the age of standardized satellites. To build a single one would be much more expensive than today's satellites because many sensors, thrusters, and logic circuits are included beyond what is needed for any particular mission. By using this design for many missions, it permits a mass production approach, reducing the unit price. Large weight and volume capacities of the shuttle relaxes the strict weight and size limitations imposed on pre-Shuttle launched spacecraft. Among the many innovations is the use of hardware which provides on-orbit exchange of components and sensors. Thus, the Shuttle can be used for maintenance and updating of these satellites.

One configuration of Spacelab, the international experimentation facility being developed by the European Space Agency, consists of a large pressurized module with an external equipment pallet. These devices will be frequent passengers on the Shuttle. Their purpose is to provide an extension of the experimenter's ground-based laboratory with the added qualities of long-term, gravity-free exposure, a location from which earth can be viewed and examined on a large scale, and a

place where the universe can be studied free of atmospheric interference. The Orbiter may be flown in an inverted attitude so that the instruments can point to earth for surveys of resources and environmental parameters. Pressure suit operations in the payload bay are practical when instrument service is needed. Each Spacelab may be flown up to 50 times over a ten year period. Crewmembers assigned to Spacelab will consist of Americans and Europeans, but the intent is that each will fly on only one mission. These people will be specially trained to carry out experiments and will be classified as payload specialists.

Before the Space Shuttle became operational, there were the series of orbital flight tests carried out between spring 1981 and mid-1982. These missions were used to evaluate the performance of the Shuttle and its systems in planned modes of operation. The Shuttle system has now been certified for operation.

The primary purpose of the first orbital flight test was to demonstrate a safe ascent and return of the Orbiter and crew. On later test flights, scientific payloads were flown to the extent allowable within the test program. The primary payload for these flights was, however, flight test instruments for the Shuttle itself. These consist of two major items, the Development Flight Instrumentation package and the Integrated Environmental Contamination Monitoring package. The first of

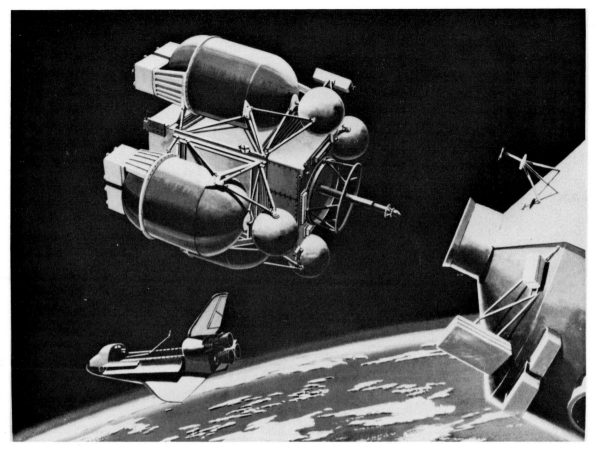

An attempt was to be made to rendezvous and dock a teleoperator with the Skylab space station. This device, after docking with the large hulk, would have either raised its orbit to extend the life of the space station or caused it to reenter into some remote ocean area. After imparting the appropriate amount of thrust, the teleoperator would have unlocked and returned to the Orbiter for future reuse in retrieving satellites. Unfortunately, delays in the program and an early reentry made the rescue attempt futile.

The Large Space Telescope will be released from the Orbiter to become a free flying observatory in the mid-1980's. There will be periodic visits by other Orbiters for repairs and servicing.

The Long Duration Exposure Facility will float freely in orbit for months at a time. A variety of experiments will be mounted on each one. This platform is reusable and offers a low-cost way to fly many kinds of instruments for scientific research.

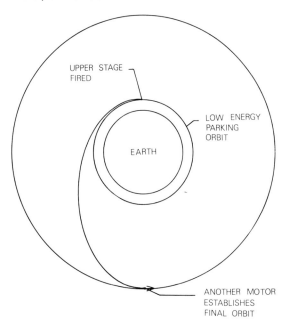

Travel beyond the Orbiter's low energy orbit required the satellites to be equipped with propulsion stages for boosting them to their final destinations. A sequence of two thrusts is necessary to raise an orbit. Only one large thrust is needed to escape the Earth for planetary flight.

these is used for measuring a variety of Shuttle performance parameters; the second is used for measuring a large number of environmental aspects related to Shuttle operations. Both of these packages were carried on every orbital flight test. They are mounted on a pallet and have a combined weight of about 10,000 pounds, using about 20 per cent of the cargo bay. Therefore, Shuttle-related scientific payloads accounted for a great deal of activity on these orbital flight tests.

On the first test mission, STS-1, Young and Crippen tried all systems and conducted many engineering tests. They checked the five computers, jet thrusters used for orientation control, and cargo bay doors allowing deployment of the radiators. (The radiators function to release heat that builds up in the crew compartment.) Still and motion pictures were taken and some were telecast to earth. One of these showed that all or part of 16 thermal tiles were lost from the two pods on the tail section. The second mission, STS-2, was planned to last at least five days. However, one of the fuel cells that converts hydrogen and oxygen into electrical power and drinking water,

The Multimission Modular Spacecraft can be used in a variety of configurations, depending on the mission and the orbit. It can be launched from the Shuttle or on an expendable booster. These satellites are designed to minimize cost through standardization and reuse.

malfunctioned soon after launch. This reduced the flight to 54 hours. Nevertheless, Engle and Truly managed to achieve most of the mission objectives. The remote manipulator arm was tested. As in STS-1, a complement of performance instruments were used to test the various Orbiter systems. At this point in the series of test missions, other types of experiments and applications were also completed as part of the mission plans. Earth surveys were gathered to provide important information for oil and gas prospecting, locating good fishing grounds, and understanding gravity effects on plant growth. STS-3 carried a large variety of experiments, too, and lasted eight days. One of these created a great deal of public interest because it was devised by an 18-year-old high school student. Following the youth's study design, Lousma and Fullerton observed the zero gravity flight behavior of bees and moths. The objective was to determine the effect of differing ratios of body mass to wing area on stability and orientation. STS-4 featured a test of an air force infrared sensor for future surveillance spacecraft. Mattingly and Hartsfield have the dis-

tinction of carrying the first military payload on a Shuttle. The first operational flight, STS-5, has four crew members, Vance Brand, Robert Overmyer, Joseph Allen, and William Lenoir. The latter two are the first mission specialists of the Shuttle program. On this flight, two other spacecraft are to be deployed, both communications satellites. One is for Satellite Business Systems Corporation and the other for Canada's Telesat. Plans for STS-6 include transporting the Tracking and Data Relay Satellite.

Let's take a look at a typical operational flight in 1985. This example mission includes a scientific payload and is seven days long. The flight consists of attaining an orbit with an inclination of 57 degrees and an altitude of 115 miles for the first day of activities. After that, the Orbiter will raise its altitude to 230 miles for the next six days to carry out special tests. The 115 mile orbit is achieved by two Orbital Maneuvering Subsystem burns. The first of these occurs shortly after external tank separation, and the second occurs about one-half hour later. The first 24 hours of the flight would be spent in a low altitude orbit to deploy the payload and to collect a variety of data under high atmospheric drag conditions at varying Orbiter attitudes. Then two more burns would be performed by the Orbital Maneuvering Subsystem rockets to place the Shuttle in a 230-mile circular orbit. Immediately following the last burn, a thermal test is to be initiated in which the Orbiter maintains an attitude so that the payload bay faces away from the earth. At the completion of this test, the Orbiter is to deploy a special target and to perform a three-hour rendezvous test. This test is required to provide information for future retrieval flights. De-orbit and landing at the Kennedy Space Center is planned to take place about 164 hours after liftoff.

COMMUNICATIONS IN SPACE

When Neil Armstrong first put his foot on the lunar surface, we saw it. Later, when two astronauts drove the Lunar Rover on the moon, we saw it. When two cosmonauts and three astronauts linked up in orbit, we saw it as it happened. These historic events were recorded and transmitted by modern communications equipment. Television is a small part of it. There were literally thousands of bits of information being sent between the Mission Control Center and each Apollo-lunar flight. This included television pictures, voice transmissions, medical data on each crewmember, telemetry on the status of most subsystems like propulsion, computers, and so forth. The Space Transportation System requires even more communications services. Not only will there be the Orbiter, but most of its payloads will be transmitting and receiving data. As we will soon see, its low orbit presents additional problems. Nevertheless, the communications network designed for the Shuttle will provide instantaneous links between the ground and the Orbiter or its payload almost all of the time. To be more specific, television and data on the subsystems, payloads, and medical parameters are sent down from orbit. Commands for the Orbiter and its payloads, as well as instructions and charts, are sent up to orbit, and voice links are, of course, set up for two-way transmissions.

The communications network consists of new Tracking and Data Relays, (i.e., two high altitude satellites working with one ground station) and the well established set of ground stations used in the Space Tracking and Data Network. The NASA communications network, which will be augmented by the ad-

dition of a domestic satellite, links the tracking stations with the ground control centers. There is also another network for payloads going into deep space, such as planetary probes.

Crewmembers on early orbital flights did not have the advantage of continuous communications with ground stations as they circled earth at a rate of once every hour and a half. There were periods of up to 20 minutes during which no one could hear from the astronauts. During the Apollo lunar flights, the situation was better because the spacecraft was so far away from earth that only a few stations were necessary to maintain links almost continuously. The Space Transportation System operates only in low-orbit where direct ground transmission coverage is very poor. Thus, the Tracking and Data Relay Satellites were specifically designed to provide almost continuous links between the Orbiter and the ground. They are also to be used with attached and free-flying payloads in low and medium altitude orbits. Nearly continuous monitoring helps to reduce the chance of experiment failure, decreases the use of onboard data storage, and allows quick changes in experiments. The two satellites in the Tracking and Data Relay System will be placed in high altitude orbits so that they remain stationary and in view of a single ground station at White Sands, New Mexico. The Orbiter and other spacecraft are to send signals to one or both of these satellites. These signals are relayed to the ground station and then sent to the Mission Control Center in Houston, Texas. Information and voice transmissions coming from Houston are sent to New Mexico via ground lines or microwaves. Then these signals are sent up through the satellites and onto the Orbiter or other spacecraft. These two satellites provide orbital communications coverage at least 85 percent of the time for all spacecraft, even for those at the lowest altitudes.

The established Space Tracking and Data Network consists of several ground stations scattered around the world for support of Orbiter launch and landing operations, as well as for propulsive upper stages and free-flying systems operating in high Earth orbit. In addition to this, the NASA communications network forms the ground links between the worldwide tracking stations and the Mission Control Center in Houston. Data from orbit will be handled in one of two ways. The Tracking and Data Relay ground station may reroute the entire amount of data through a

One Spacelab configuration consists of a large pressurized module for a team of researchers and external equipment pallets. Among the instruments will be telescopes, long booms for collecting data on charged particles and electromagnetic waves, and communications experiments.

Western Union will provide communications services to and from the Space Shuttle via the Tracking and Data Relay Satellite System. There will be two such satellites in high orbit over the Earth. A single ground terminal is located at White Sands, New Mexico.

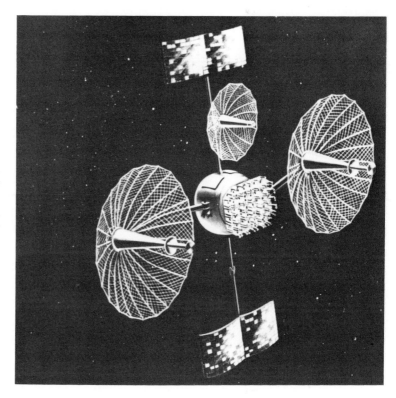

Satellite tracking antennas typically have large dishes which must be steered to point at the spacecraft. This one is 85 feet in diameter and is located at Fairbanks, Alaska.

domestic communications satellite to an associated ground terminal at the Johnson Space Center. Any agency in the continental United States having a domestic satellite terminal will also have access to this data. The other approach to handling this is to selectively channel the data. In such a situation, the Tracking and Data Relay ground station would route selective portions of the data through a circuit directly to the Mission Control Center.

This vast communications network and facilities relieve the workload on the Orbiter crew. In fact, during operation of attached payloads and checkout of free-flying satellites, the two pilots can take time to catch up on their reading. Even the mission specialist can relax during many experiment periods. After all, the Space Transportation System will have a regular staff of pilots and mission specialists who may fly several times each year. They will be good at their jobs and budget their working hours wisely.

MISSION CONTROL CENTER

The Mission Control Center is located at the Lyndon B. Johnson Space Center just

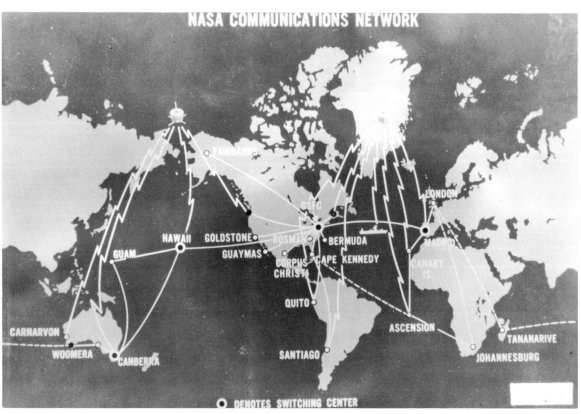

The NASA Communications Network will assist in Space Shuttle communications. A domestic satellite will be added to the network to further enhance its usefulness.

The Lyndon B. Johnson Space Center (JSC) is located just south of Houston, Texas. It is the home of the Mission Control Center and headquarters for the astronaut corps.

The Mission Control Center is the brain center for space flights which carry crews. Here we see the level of activity associated with the first free-flight of the Enterprise on August 12, 1977.

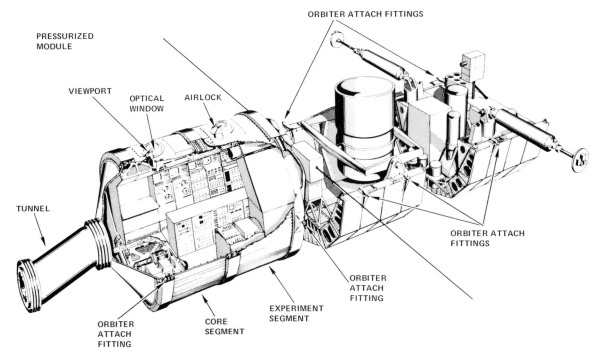

This very sophisticated configuration for the Spacelab experiment platform consists of a large pressurized module in which a crew of researchers can work in a shirt-sleeve environment. Two external pallets are shown here, one with a telescope and one with several data collecting instruments for measuring electromagnetic waves and charged particle densities. During the experiments, two large booms would be extended beyond the Shuttle for collecting certain types of data. This whole configuration fits within the Shuttle payload bay and remains there during the entire flight. The tunnel is used for access to and from the Shuttle crew cabin.

south of Houston, Texas. This center has been designated as the Shuttle operator for all NASA flights. The Mission Control Center provides monitoring and emergency support, two-way communications with the Orbiter, and collection of flight data at a central point for all flights, both NASA's and other users'. This facility may be thought of as the brain center for space flights which carry crews. Visitors are sometimes permitted in a small booth which overlooks the rows of control consoles and TV monitors staffed by dozens of ground controllers, each with a separate and vital job. This is the same facility that was used for the Apollo lunar flights and the Skylab missions.

Each of the ground controllers in the Mission Control Center has specific instructions and training for his or her assignment. Before each flight, all planning is carried out in detail. Persons in the control center are issued copies of a book containing instructions for almost any situation that might arise during a particular mission. This flight control team is on duty 24 hours a day for the duration of each mission. They work in shifts when all is well.

In cases of emergency, many work around the clock with only an occasional break for sleep and nourishment. They are there during launch, orbital operations, and re-entry. When Columbia touched down on the runway after that first orbital flight, you could hear a loud cheer from Houston.

During the Apollo-13 lunar mission, an explosion occurred while en route to the moon. From the moment Jim Lovell transmitted "Houston, we've got a problem" to the Mission Control Center, the team worked feverishly to save the crew. Nobody relaxed until splashdown, several days after the accident occurred. The ground controllers literally had to rewrite the mission plan on the spot. A plan that had taken years to develop was changed in hours for a single purpose, to bring the three crewmen back to earth safely. They did just that.

REGULAR SHUTTLE TRAVELLERS

The Space Transportation System will carry a number of repeat riders such as Spacelab, a variety of upper stages which are attached to satellites and planetary probes, the

Long Duration Exposure Facility, and the Multimission Modular Spacecraft. Let's briefly review the configuration and purpose of these regular travellers to the cosmos. Spacelab is a versatile laboratory for staffed and automated activities in near-earth orbit. Its primary objective is to provide the scientific community with easy, economical access to space. This laboratory is being built in Europe with European funds. It is carried by the Shuttle and remains attached to the Orbiter during all phases of the flight. Spacelab generally consists of a module and pallet sections arranged in various ways to suit the needs of a particular experimenter. The pressurized module is accessible from the Orbiter crewcabin through a transfer tunnel and provides a shirt-sleeve working environment. This module is composed of one or two short cylinder segments, each 13 feet-4 inches in diameter and 8 feet-7 inches long. The pallets accommodate experiment equipment for direct exposure to the space environment. A standard pallet section is 9 feet-10 inches long. These can be connected in a series of two or more in the cargo bay. If no module is to be used, a cylindrical *igloo* can be mounted on the end of the first pallet. This provides a controlled and pressurized environment for a few critical Spacelab systems normally carried in the module. If only the pallet is to be used on a particular flight, all experiments are operated remotely from the aft deck of the crewcabin or from the ground.

A total of ten basic flight configurations have been developed to meet almost all user needs. There is the regular module without pallets, and there is the regular module with one or two pallets. The components can also be arranged with the short module and two pallets or the short module and three pallets. Also possible is the configuration using one, two, or three pallets, and the one with four pallets only. Finally, there is the configuration with five pallets all in a row. Although the Shuttle can carry up to a maximum of 65,000 pounds, it is only allowed to bring back 32,000 pounds from orbit. Therefore, the maximum weight of a Spacelab is to be limited to 32,000 pounds because it will remain attached to the Orbiter. In general, Spacelab missions will concentrate on intense, short-term investigations. Thus, it will complement those long-term observation programs that use free-flying satellites. Missions are designed to provide the greatest scientific return from each flight. The

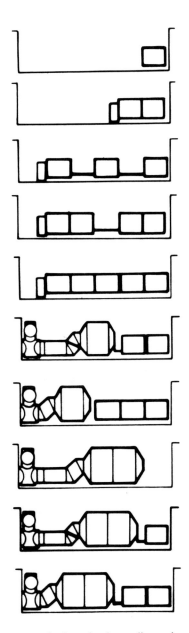

There are a total of ten basic configurations of the Spacelab experiment base. You can have one pallet, two pallets, three pallets, or all the way up to five pallets without a module. For the more costly programs, you can add a staffed module, either the short or regular size. Usually, a module has one or more pallets in which vacuum-type experiments are mounted.

payload specialist, trained by the user and working with ground based scientists, is an integral part of each flight.

A variety of scientific operations will be typical Spacelab missions. Astrophysics experiments will investigate a wide range of long term scientific problems, including origin and future of matter, nature of the universe, life cycle of the sun and stars, and evolution of the solar system. Solar physics payloads involve instruments designed to obtain data which will allow us to understand the physical processes of energy production within the solar interior. We should also be able to learn more about the transport of energy through the solar atmosphere and its ultimate dissipation through radiation and solar wind. The emphasis of early missions will be on the interaction between the sun and earth processes. A dedicated Spacelab mission will be used to study the atmosphere and magnetic field about the earth. There is also an advanced technology laboratory on one of the Spacelab flights which will emphasize developments in technology. There will be flights which investigate the movement of continents and dynamics of our oceans, pollution in our atmosphere, and changes in our weather and climates. Numerous experiments will be carried out in the field of life sciences. These will extend our knowledge in the areas of medicine and biology, allowing us to better deal with our earth-bound needs. There will be many flight opportunities at a relatively low cost, thus permitting wide participation from the public and private sectors. Experimenters will be able to fly with their experiments and use existing or off-the-shelf hardware.

Initially, seven-day missions are scheduled for the life sciences experiments. Eventually, dedicated flights as long as thirty days are envisioned. For a given flight, ten to twenty life sciences experiments will be carefully selected and developed that can be operated by one or two onboard payload specialists. They will have television, voice, and data links with ground-based scientists and engineers to help them. Several dedicated life sciences missions are already being developed.

The expendable upper stage is a simple, low-cost device used with spacecraft going to different altitudes and inclinations beyond those of the basic Shuttle capability. Upper stage systems consist of one or more solid-propellant propulsive units, electronics, and special communications equipment. Two types of upper stages have been developed for the Space Transportation System. The first is a solid propellant, spin-stabilized unit, called the spinning *solid upper stage*. There are two sizes, one that can send two thousand pounds to high altitude and one that can send four thousand pounds to a high altitude. The com-

mon characteristic of these spinning stages is that they do use spin stabilization techniques during the ascent phase of the flight. By causing a spacecraft to spin on release, there is less likelihood of it tumbling or moving out of the desired path of flight. The other type of stage is also a solid propellant device but it uses three-axis stabilization and is called the *inertial upper stage*. It was intended to come in a variety of combinations of motor sizes and can boost either a single or multiple spacecraft to higher orbits or even into planetary trajectories. What keeps spacecraft using this system stable are mechanisms within the spacecraft itself which are operated by external control. The spinning stages are primarily intended for use during the transition period between 1981 and 1985 while satellites are designed for both expendable boosters and the Space Shuttle. For example, the international communications satellites built for the late 1970s and 1980s will be launched both on Atlas-Centaur boosters and in the cargo bay of the Shuttle. One spacecraft design is being used for all of these launches. A spinning upper stage will be used when launched in a Shuttle in order to take the place of an upper stage of the expendable booster. The Orbiter performs the initial pointing, spin up, and release of the satellite-with-spinning-upper-stage. These procedures are initiated and controlled from the aft flight deck. Payloads may be spun up to as much as 100 rpm before release. After the combination payload is released, the Orbiter maneuvers to a safe distance. The payload and Orbiter coast in that parking orbit for approximately 45 minutes until the appropriate crossing of the equator. At that time, the staging motor is fired and injects the spacecraft into a transfer orbit which will take it to an altitude of 22,200 miles. As it reaches this altitude, a solid rocket motor within the satellite itself is fired to inject the spacecraft into its final high altitude orbit. It then joins many other satellites in that orbit and becomes an operational communications satellite.

The inertial upper stage, under development by the U.S. Department of Defense, relies on a three-axis control system which uses propulsive devices to maintain its orientation. Its job is to place a larger class of spacecraft or multiple spacecraft into orbits or trajectories not within the Shuttle capability. The inertial upper stage/spacecraft combination is deployed from the payload bay by means of the remote manipulator arm. The inertial upper stage family consists of two basic two-stage vehicles with a three- or four-stage configuration for the higher energy mission. Only the smallest version has been developed to date. All stages consist of solid rocket motors as well as supporting structure and sub-systems needed for control and sequencing.

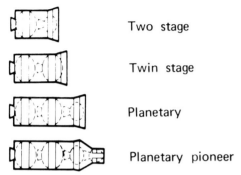

Two stage

Twin stage

Planetary

Planetary pioneer

The Inertial Upper Stage has four basic configurations and used primarily for boosting large payloads into higher orbit or for injecting spacecraft onto interplanetary trajectories. There is the two-stage configuration which is the smallest version and is the only configuration developed so far. Then the twin-stage in next in size, followed by the planetary configuration which is a three-motor configuration. Finally, the largest is called the planetary pioneer and consists of four solid rocket motors. Each configuration also has a sophisticated guidance and control package which keeps it inertially oriented in the proper direction.

The *Long Duration Exposure Facility* is being developed by NASA as a reusable, passively-stabilized structure which can accommodate many technology, science, and applications experiments. The only requirement for placing a payload on this device is that it requires exposure to the space environment. It should provide an easy and economical means for conducting such experiments. The ranks of users are expected to include governments, universities, and industries in both the United States and abroad. This exposure facility will be delivered to earth-orbit by the Shuttle where it will remain for at least six months at a time before being retrieved and returned. Upon arrival in orbit, the remote manipulator arm will remove this device with its mounted experiments from the cargo bay and point its long axis toward earth. Upon release this orientation will allow the long duration exposure facility to remain in a passive orientation as it circles earth each 93 minutes at an altitude of about 345 miles.

During a six-month period its altitude will decay about 23 miles due to exposure to the upper atmosphere. The long duration exposure facility is about 30 feet long and has room for 72 experiment trays on the drum-like structural framework. Each tray is approximately 50 inches long and 38 inches wide and may have depths of 3, 6, or 12 inches. The total mass allowable in a single tray is 175 pounds. If you can think of any experiments

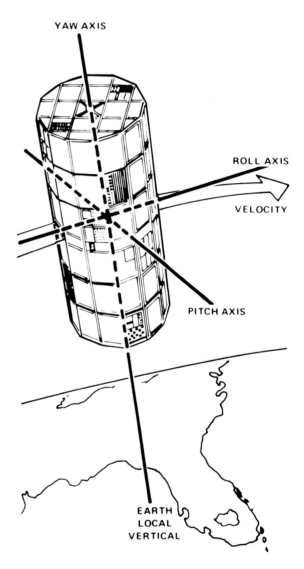

*The Long Duration Exposure Facility will naturally take an orientation with its long axis pointing toward the Earth. This is known as a gravity-gradient stabilized configuration. It will remain in **this orientation and permit the many experiments** mounted about its structure to be exposed to the environment of space for many months at a time. At the end of each mission they will be retrieved by the Shuttle. The users are expected to include governments, universities, and industries in both the United States and abroad.*

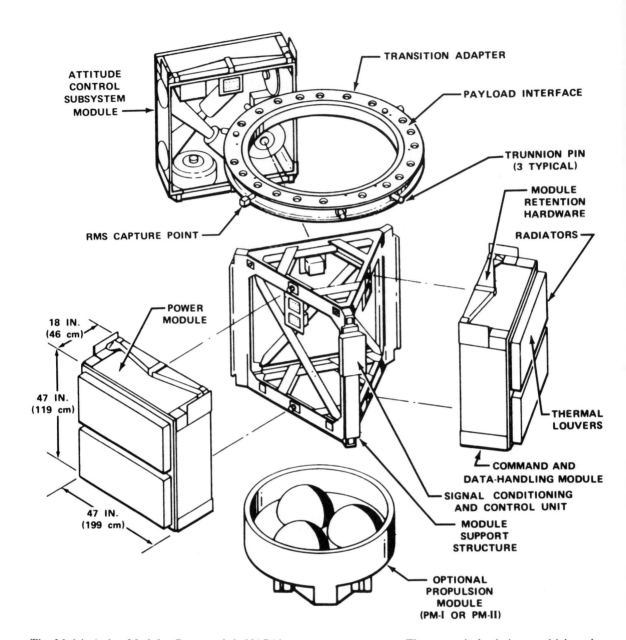

ATTITUDE CONTROL SUBSYSTEM MODULE

TRANSITION ADAPTER

PAYLOAD INTERFACE

TRUNNION PIN (3 TYPICAL)

MODULE RETENTION HARDWARE

RADIATORS

RMS CAPTURE POINT

POWER MODULE

18 IN. (46 cm)

47 IN. (119 cm)

47 IN. (199 cm)

THERMAL LOUVERS

COMMAND AND DATA-HANDLING MODULE

SIGNAL CONDITIONING AND CONTROL UNIT

MODULE SUPPORT STRUCTURE

OPTIONAL PROPULSION MODULE (PM-I OR PM-II)

The Multimission Modular Spacecraft is NASA's space-age erector set. There are six basic items which make up a typical spacecraft. The payload package is mounted to this modular configuration. If a great deal of maneuvering is required, then the optional propulsion module is included in the spacecraft package.

which can be contained in this volume and weight and are designed to measure some property of the space exposure, then the long duration exposure facility is the right vehicle to carry this into space.

The *Multimission Modular Spacecraft* is designed to be used in a variety of Earth orbits from low to high altitudes. It is particularly useful for remote sensing missions. Many of these spacecraft with their payloads will either be brought back from space or serviced on-orbit by the Shuttle. The basic modular spacecraft is made up of six items. These are the modular support structure, the transition

adapter, the command and data handling subsystem, the attitude control subsystem, the modular power subsystem, and the optional propulsion module. You might think of this as NASA's space-age erector set. This is, in fact, the intent. By taking these building blocks and putting them together to make a variety of spacecraft, a major cost savings will be realized.

As we can now see, the Space Transportation System is much like a freight train to orbit. It can carry a variety of regular cargo plus a number of specialized spacecraft. Its unique capabilities have led to a new breed of

A prototype of the 50-foot-long manipulator arm is shown here moving a 60-foot inflatable payload during tests at the Johnson Space Center.

travellers: Spacelab, expendable upper stages, the long duration exposure facility, and the multimission modular spacecraft. None of these were practical or possible before. More regular travellers will surely appear in the next few years.

EXTRAVEHICULAR ACTIVITIES

If you have a flat tire while driving down the road, you simply pull over and change it. This may be thought of as a form of extravehicular activity, since you had to get out of the vehicle to make repairs. If an airliner experiences a malfunction while flying en route, it lands at the next convenient airport and repairs are made. An extravehicular activity is, of course, not possible while flying the airliner. If an Orbiter has a problem that must be taken care of outside the crew cabin while circling the earth, they cannot just pull over or land and make repairs. It may well have to be fixed while in orbit. A crewmember who ventures outside the confines of the comfortable cabin to carry out some function is said to be on an *extravehicular activity*. Every Shuttle flight will carry equipment for activities such as crew rescue from a disabled Orbiter, inflight inspection and repair, payload servicing, and so on. In fact, sufficient supplies are provided to perform three separate extravehicular activities for two crewmembers on each mission. Thus, there are two sets of spacesuits and associated equipment on each Orbiter. Two of these three ventures to the outside environment are intended to be used for payload-related activities which are planned or unscheduled but are not related to emergencies. The third is reserved for any possible contingency situation.

The Gemini Program helped provide initial indications about the potential of being able to leave the spacecraft. The Apollo, lunar, and earth orbital extravehicular activities applied this capability to performing required duties about the huge space vehicle. The Skylab program, in addition to using extravehicular activities as part of the original flight plan, used them for system repairs essential to save the mission because of a mishap during launch of the vehicle itself. Fortunately, repairs worked out successfully, and the Skylab mission reaped many benefits as our first space station. The Shuttle program is providing accommodations to take advantage of the great maneuverability provided by this type of activity. To give you some examples of the numerous variety of payload related tasks that can be carried out, the following list is provided:

1. Inspection and photography;
2. Installation, removal, and transfer of film cassettes, material samples, protective covers, and instrumentation;
3. Operation of equipment, including standard or special tools, cameras, and cleaning devices;
4. Cleaning of optical surfaces;
5. Limited connection, disconnection, and storage of fluid and electrical umbilicals;
6. Replacement and inspection of modular equipment and instrumentation on the payload or spacecraft;
7. Repair and repositioning of antennas and solar arrays;
8. Conduct extravehicular experiments;
9. Provide mobility outside the payload bay and in the vicinity of the Orbiter using a Manned Maneuvering Unit;
10. Mechanical extension, retraction, or jettison of experiment booms;
11. Handling of Shuttle passenger personnel rescue systems in cases of transfer from a disabled Orbiter to the rescue spacecraft; and
12. Transfer of cargo, including incapacitated crewmembers.

Extravehicular activity has been the term used to describe those activities performed by astronauts outside the spacecraft in the environment of deep space. Spacesuits and portable life support systems have been designed to provide the human body with sufficient protection from the extremes of temperature and vacuum pressure. In the Shuttle program extravehicular activities are defined as all activities which require a crewmember to don a spacesuit and the life support systems and leave the Orbiter crew cabin. This includes operations outside the payload bay in the vicinity of the Orbiter as well as those performed in the payload bay with either the doors opened or closed. *Intravehicular activities* are considered those activities which take place within the cabin of the Orbiter or the Spacelab and do not require spacesuits. Three basic classes of extravehicular activities have been identified which cover all currently defined related activities: those that are planned, unscheduled, or emergency. A *planned* activity refers to tasks planned prior to launch in support of payload operations.*Un-*

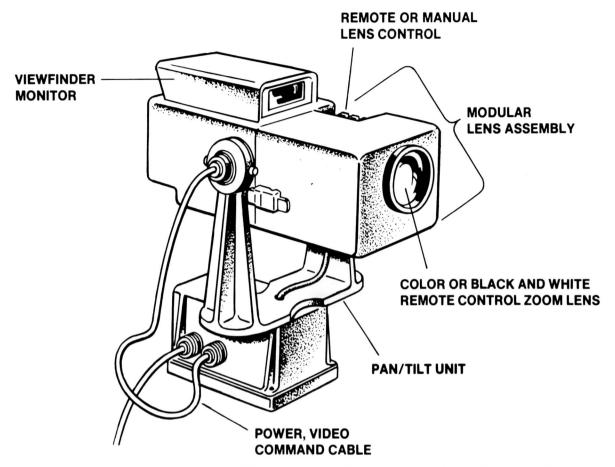

REMOTE OR MANUAL
LENS CONTROL

VIEWFINDER
MONITOR

MODULAR
LENS ASSEMBLY

COLOR OR BLACK AND WHITE
REMOTE CONTROL ZOOM LENS

PAN/TILT UNIT

POWER, VIDEO
COMMAND CABLE

Many remotely controlled T V cameras will be located in the Orbiter cargo bay. Each will be controlled from the aft crew section within the cabin.

scheduled activities are those that are not planned, but which may be required to achieve or enhance mission success. Finally, the *emergency* extravehicular activity covers all those external tasks required to safely return the Orbiter and its crew.

The consumables or supplies needed for an external activity include nitrogen and oxygen required for repressurizing the airlock and oxygen and water for charging the spacesuit backpack. Prior to leaving the Orbiter on an external activity, the crewmember will prebreathe oxygen to eliminate nitrogen from the blood stream in order to minimize the possibility of experiencing the bends. This could occur because the spacesuit pressure is only one quarter of that in the crew cabin. Almost pure oxygen must be breathed for a period of three hours before exposure to a lower pressure.

While part of the crew is floating around outside, continuous communications are essential. The Orbiter provides such com-

munications between crewmembers inside and outside the cabin. Ground controllers can also communicate with all crewmembers at this time, via the Tracking and Data Relay Satellite system. Voice transmissions and medical data on external crewmembers are being transmitted to the ground continuously. In this way, if any extravehicular astronaut has a medical problem, it will be detected immediately.

In fact, there is such a variety of communications equipment on the Orbiter, you would think the place was bugged. There is an intercom system within the Orbiter crew cabin, and oxygen masks worn during prebreathing are fitted with microphones and headsets. The crewmembers leaving on an extravehicular activity will have communications in the airlock while donning their spacesuits. Once their helmets are on, they will be able to turn on the radio in the backpack, and voice and data will be transmitted to an antenna in the airlock itself. This communica-

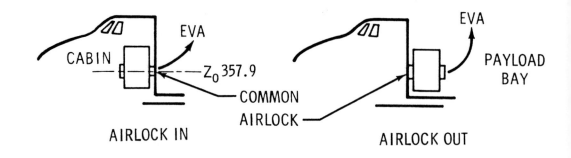

AIRLOCK IN   AIRLOCK OUT

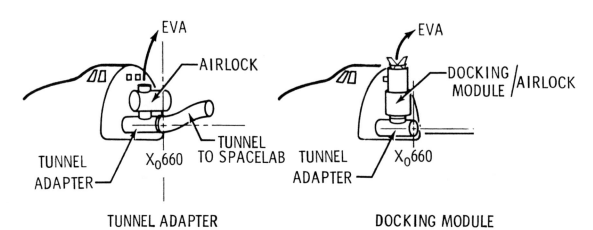

TUNNEL ADAPTER   DOCKING MODULE

The Orbiter airlock can be positioned in one of three basic ways. Within the crew cabin on the mid-deck, it is oriented in a vertical position. If it is placed in the payload bay, it can be either vertical or horizontal in its orientation. It may also have an adapter for the Spacelab tunnel or for docking with another vehicle.

tions mode continues until the crew leaves the airlock at which time transmissions will be picked up by an outside antenna.

A closed-circuit television system is used onboard the Orbiter for monitoring the payload bay area. This system permits easy use of the remote manipulator arm and provides a means of overseeing extra-vehicular activities in the bay itself. The system has the capability to transmit color television pictures to the ground. Thus, you and I will be able to watch the activities right at home. In addition, there are two black and white television monitors in the cabin located at the aft crew station area near the remote manipulator controls. There is a handheld color television and monitor provided in the cabin for coverage of inside operations. There may be as many as three black and white cameras in the payload bay, located forward and aft and, if required, in the lower part of the bay. These cameras contain pan and tilt mechanisms to position them for better coverage of operations. Each remote manipulator arm has a black and white camera with

pan, tilt, and lens controls. This will permit manipulations of the arm which are out of normal view and will allow inspection of areas not visible by other means. The control unit for pan and tilt mechanisms is located at an aft crew station.

Just below the flight deck is the mid-deck. This area is primarily the crew living and equipment storage area. As you will recall, it contains the airlock and exit path to the payload bay. Preparation of equipment for external activities is accomplished on this level. An *airlock* is a device used to transition from the shirt sleeve cabin environment to the zero pressure of the space vacuum. Before launch, the Shuttle airlock can be positioned in one of three ways: vertically in the cabin mid-deck, vertically in the payload bay, or horizontally in the payload bay using a tunnel adaptor in order to be more compatible with the various payload configurations. The inside dimensions of the airlock are 63 inches in diameter and 83 inches in height, and a 40-inch D-shaped hatch is provided as an exit into the payload bay. The airlock also

The Manned Maneuvering Unit permits an astronaut to maneuver close to a free-flying payload either to make repairs or to retrieve it. The Orbiter station remains nearby the spacecraft while the astronaut is doing his job.

provides depressurizing and repressurizing functions to allow crew transfer to and from the payload bay. In addition, it provides a storage area for spacesuits and other related equipment. Each of the hatches on the airlock contains a small viewport that allows the crewmembers to be observed while in the airlock. This small window has a minimum diameter of four inches. Spacesuits and other equipment are put on in the airlock itself. At the completion of suit-donning the crewmembers will have installed all the required personal gear with the exceptions of the helmets and gloves which are left off until the start of checkout of the life support system. Handholds and foot restraints are provided in the airlock to facilitate dressing and undressing.

The Manned Maneuvering Unit gives the external crewmember the capability to reach areas outside the payload bay without the need for handholds or other aids. This is a self-contained unit which assists by providing propulsive forces and will allow crewmembers access to areas such as the Thermal Protection System and the aft end of the Orbiter not normally reachable by the remote manipulator arm.

The Manned Maneuvering Unit is a modular device which is easily attached to the spacesuit when needed. It is stowed in the payload bay at a support station, making it extremely convenient for getting it on and taking it off. A maximum of two such units can be carried on each mission. Non-contaminating gaseous nitrogen is used as the propellant. Its nominal operating range from the Orbiter is 340 feet. It even has two electrical outlets which provide power for operating movie or television cameras, floodlights, or power tools. The unit itself weighs about 230 pounds and it takes about 15 minutes to put it on and check its operation. It can be restowed in about the same amount of time.

In a rescue situation, the maneuvering unit provides the most practical means for retrieving crewmembers from an unstabilized or tumbling, disabled Orbiter. For payload support, this unit will allow access to instruments on the ends of long booms or appendages which extend outside the payload bay beyond the reach of the manipulator arm. For free-flying payloads, this maneuvering unit could allow retrieval of small objects and access to large vehicles that are sensitive to Orbiter thruster perturbations and contaminations. In such a situation, the Orbiter would station-keep (or park) at a safe distance, and the extravehicular crewmember would maneuver to the payload and retrieve contamination-sensitive instruments or install protective covers. Eventually these maneuvering units will be essential to the assembly of large structures in space.

The spacesuit with its life support backpack provides protection and life supplies for a crewmember for up to six hours in duration. Its purpose is essentially the same as that of the Apollo spacesuits. However, it is a new design, tailored for the Shuttle era of extravehicular operations. The suit is unique in

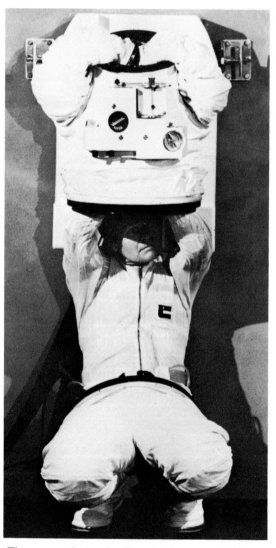

The spacesuit for the Shuttle era is a new design. The major portion consists of two parts, the upper torso and the lower torso. In addition, there are a pair of gloves and a helmet. To don the suit, first you put on the lower torso, then snuggle into the upper torso while it hangs on the wall of the airlock.

The crewmember who is about to make an extravehicular activity completes the donning of the spacesuit by checking out the system after the helmet and gloves have been secured properly. The helmet is a transparent bubble that has an add-on visor assembly which has side-shades as well as a center-shade for protection against solar radiation. The backpack is already attached to the upper torso when the spacesuit is donnd. After all checkouts are completed, the astronaut is ready to walk in space.

that it has a minimum of components and needs very few connections and disconnections by the crewmember while putting it on and taking it off. Many of the hoses and straps that were seen in the past have been eliminated and controls and displays are easily accessible. A unique torso design allows the elimination of costly customized suits. I would

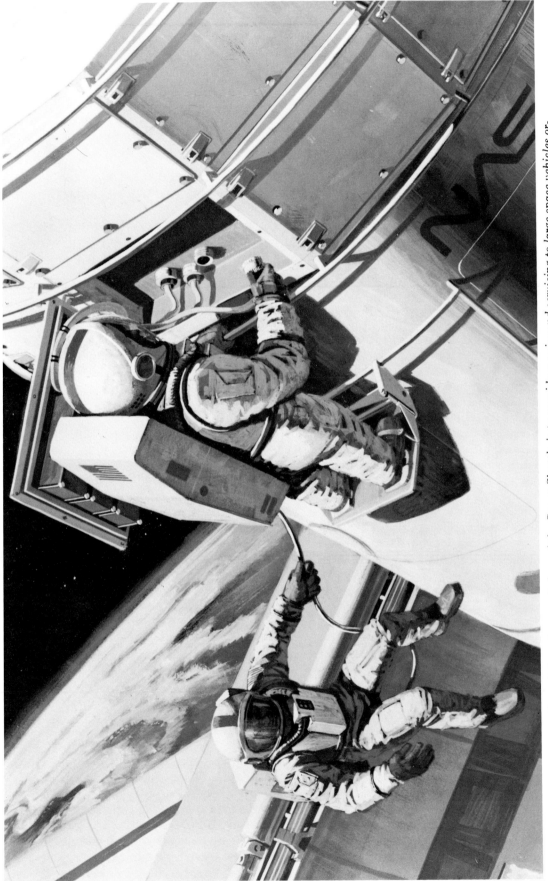

One of the primary missions of the Space Shuttle is to provide repairs and servicing to large space vehicles orbiting the Earth. Here we see two extravehicular crewmembers working on a large space telescope repair job. This is not your typical line crew.

not say that "one size fits all," but certainly one size will fit most of the crewmembers who will be using them. This new suit with its integrated backpack weighs about 182 pounds. The suit itself consists of a pressure garment, a cooling garment, and a helmet assembly. The life support system includes oxygen, water for cooling, and communications equipment. It, of course, also has a variety of other support equipment.

The pressure garment consists of several items: hard upper torso, lower torso, helmet, pair of gloves, extravehicular visor assembly, communications carrier assembly, urine collection device, and drink bag. It has fabric joints and sealed bearings at the shoulder, arm, and wrist. Arms and legs are adjustable, and standard sizing of boots is used with provision for the Skylab type foot restraints. Thermal and micro-meteoroid protection is integrated in the design. The helmet is a transparent bubble that interfaces with the visor assembly. This visor has center and side shades and provides thermal and solar radiation protection. Located within the helmet is the communications carrier assembly, a fabric headgear that contains built-in microphone and earphone electronic modules. Personal sanitation needs are met with the urine collection device in the form of a flexible container, worn by the crewmember inside the suit. This piece of equipment retains all urine collected during suited operations. With the completion of the extravehicular activities, the contents of this container are transferred into the Orbiter waste management system. Also included in the space suit is an insuit drink bag, designed as a flexible pouch which is installed in the upper torso area. It contains water for drinking during outside activities and has a drink tube and a valved mouthpiece for ease of use.

NASA has made extensive studies of the possible ways that the Orbiter might malfunction while in orbit. In those cases where crew rescue is required, it was concluded that an extravehicular activity would be the best way to accomplish this. This would entail the transfer of crewmembers from one disabled Orbiter to a rescuing spacecraft. The extravehicular activity has been established as the primary mode of crew rescue except on those missions where docking between Orbiters is possible. In these cases, a shirtsleeve environment would be provided for the transfer between the vehicles. However, where a docking module is not available, the rescue equipment would consist of spacesuits for two crewmembers and Personnel Rescue System units for the remaining people. These Personnel Rescue units are incredibly simple and, in fact, look like large-size cocoons. Each is self-contained and provides life support for up to one hour for a single crewmember. They are designed simply to contain a single person in a pressurized atmosphere for the purpose of transfer to the rescue Orbiter. It consists of a spherical pressure enclosure that provides thermal and micro-meteorite protection, a water cooled vest, and a portable, oxygen rebreathing system. It weighs 24 pounds and incorporates a viewing port.

Let's quickly run through all the events associated with an extravehicular activity. A three-hour period of breathing oxygen is required before being exposed to lower pressure. During this three-hour period a crewmember will put on a prebreathe mask and continue normal activities for a period of two hours. In the third hour the person will unstow and prepare equipment for the extravehicular activity. Upon entering the airlock, the spacesuit will be put on. At the end of three hours the prebreathe mask will be removed and helmets and visors will be donned. It takes about three-and-one-half hours from the start of prebreathing until depressurization of the airlock and opening of the hatch to the outside environment. Extravehicular activities can be performed outside for up to six hours. This time period includes checking out the Manned Maneuvering Unit and carrying out all the tasks that are required. At the end of this six-hour period, each crewmember will again enter the airlock and close the hatch. Repressurization then takes place, followed by the removal of the helmet and gloves. The spacesuit is completely removed and recharged before stowing so that it is ready for the next extravehicular activity. The total time from the start of prebreathing until all activities have ended is about 11 hours. No breathing of pure oxygen is required following the repressurizing because the danger of the bends only occurs on depressurization. Thus, each extravehicular activity entails a full day of work.

NOT ALL THINGS TO ALL PEOPLE

The Space Transportation System will undoubtedly be a boon to earth orbital activities. You have now learned a great deal about its capacity to carry large payloads into orbit and

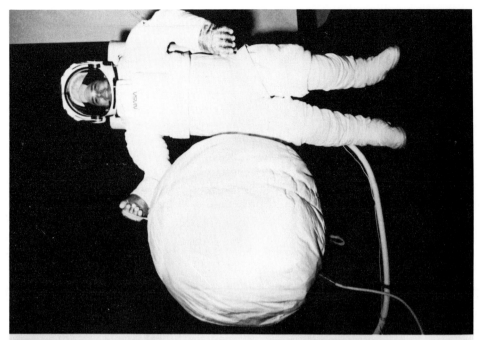

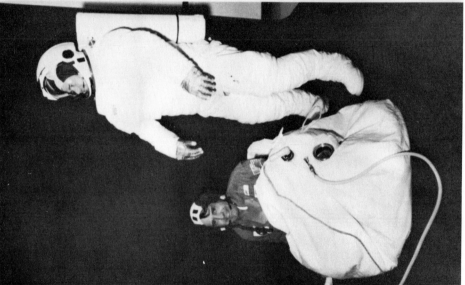

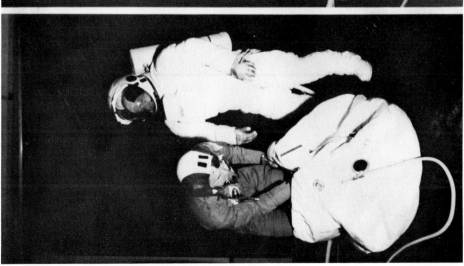

Engineers at the Johnson Space Center have developed a new spacesuit and rescue system for use by crewmembers on the Space Shuttle. Two astronauts will be outfitted with spacesuits while the other Shuttle passengers will be provided with a Personnel Rescue System. In the event an Orbiter becomes disabled and its crew must be rescued, another Orbiter will be launched, and passengers will be transferred in these enclosures. Each one has a spherical shape with a 34-inch diameter. They provide life support for a short period of time and communication systems.

its versatility relative to the types of things that will be carried out. Nevertheless, the Shuttle will not fill all needs in space. It has many limitations and restrictions. For example, let's consider the maneuverability of the Orbiter. The initial orbit for all payloads will be limited to a narrow range of altitudes that the Shuttle can reach. These are referred to as *low energy* orbits. All expendable launch vehicles now place payloads into this type of low circular orbit initially. Destination beyond the capability of the Space Shuttle will be attained through the use of expendable upper stages. However, these stages for Shuttle payloads represent added cost beyond the basic fares. In the case of expendable launch vehicles, there is no additional cost associated with injecting payloads into high-altitude transfer trajectories. Earth observation satellites must be placed in the proper orbits to pass over land or water areas of interest at a given time of day. Pre-selection of the orbit would not be particularly critical if the Orbiter had unlimited maneuverability in space. The large mass of the Orbiter, coupled with its limited fuel carrying capacity, limits the Shuttle, for all practical purposes, to the orbital plane it attains upon entry into space. Thus, its inclination and celestial longitude are fixed for all intents and purposes. Small changes in altitude are possible, however. Furthermore, the Orbiter cannot carry maximum weight to all desired orbits. Additional height can be achieved by adding up to three extra sets of propellant tanks for the Orbital Maneuvering Subsystem engines. On the other hand, the weight of this propellant can add up to 42,000 pounds which must be subtracted from the maximum payload capability.

It is true that the capacity of the Shuttle could be used to carry heavy objects into low energy orbits or to carry combined payloads on a single flight to reduce each individual user's launch costs. However, surveys show that potential users do not consider the ability to combine flights an advantage because it presents a number of schedule, interference, and risk problems. For example, what do you do if another satellite malfunctions, leaking all over yours during ascent to orbit? Another situation which might arise is one in which another user's payload is supposed to leave the Shuttle first but delays occur to the extent that your payload could not be released within the given amount of time necessary for a successful mission. These are real concerns to potential Shuttle users. Finding payloads going to the same orbital plane is not always easy because many payloads need to be launched into specific orbital planes to be effective. Unless payloads on the same flight are going to virtually the same orbital plane, upper stages or some other form of propulsion are needed. Using upper stages or other devices also increases the user's cost, partially reducing the savings of combining payloads.

NASA administrators have stated that one of the big advantages of the Space Transportation System is that payloads will be checked-out before release into orbit. If something does not work, it could be serviced or returned for repair. In the past, some satellites have completely failed shortly after insertion into orbit. The usefulness of on-orbit checkout is uncertain. Over the years, increased reliability of electronic circuitry has reduced the incidence of early satellite failure. A recent survey of 57 satellites showed that only five have had unsuccessful missions due to first-day failures. Satellite malfunctions are now attributable primarily to wearout due to interruption of power supplies because of the limited number of battery recharges or the exhaustion of propellants that maintain a satellite's stability or position. Thus, on-orbit checkout may be impractical, except for a few payloads. Furthermore, there may be little need to checkout satellites which will be subjected to the stresses of upper stages before finally reaching their operational altitudes.

Retrieval which enables payload refurbishment and reuse is a key factor in justifying the development of the Space Transportation System. It also has been a controversial issue. In 1973 the National Academy of Sciences concluded that payload recovery and in-orbit servicing would be economical for expensive systems, such as orbiting observatories, but for the less expensive payloads, the economic advantages are unclear. In another survey, a number of potential Shuttle users were interviewed on this subject. All believed the demonstrated reliability of current satellites was quite satisfactory. As on-orbit lifetimes are now approaching seven years, and may reach ten years soon, returning a satellite that incorporates antiquated technology would not be particularly useful. However, NASA is developing the Multimission Modular Spacecraft to resolve some of the cost and technology questions in this area.

Crewmembers having to be transported in the Personnel Rescue System during an on-orbit transfer between two Orbiters will be somewhat cramped in their accommodations. Nevertheless, this represents a simple and effective way of transporting personnel in cases of emergency. The plastic sphere shown here is used to depict the available volume for a crewmember.

Retrieval is impossible for objects beyond the Orbiter's limited range because expendable upper stages do not have retrieval capability. Even if the Orbiter can maneuver within the range, an out-of-control satellite cannot be recovered. Either the satellite's stabilization system must be working and capable of being deactivated, or the Orbiter must carry equipment capable of neutralizing the motion. Many satellites are oriented by means of gyroscopes which are rotating wheels that resist changes in orientation. If such an active satellite were pulled into the Orbiter using the manipulator arm, the resistances could set up severe stresses and damage the arm. A satellite is safe to recover only if all internal and external movements are neutralized.

We have now seen what the Orbiter can do and what it cannot do. We have gone through pre-launch activities at the Kennedy Space Center, have lifted off into space, and have seen what is involved in operations on-orbit. With the completion of on-orbit activities, the crew looks forward to returning home. We are about to experience this exciting and eventful phase of flight, *return to earth.*

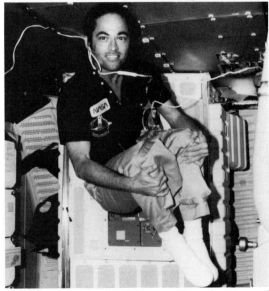

Astronaut Robert L. Crippen takes advantage of zero gravity on STS-1 to do some rare acrobatics aboard Columbia.

Initiation of reentry requires a retro-fire burn about halfway around the world from the landing site. To do this the Orbiter must fly backwards so that its maneuvering engines point in the direction of flight.

Large space structures appear to be in our future. This is an artist's concept of construction activities in low Earth orbit. When all work is done, this will become a solar power station and raised to an altitude of over 22,000 miles above the Earth.

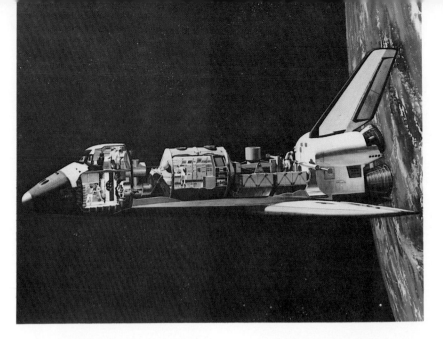

Astronauts and Payload Specialists busy themselves with a variety of tasks aboard a Spacelab-equipped Shuttle. This cutaway view shows a crew of seven doing a number of things including an extravehicular activity.

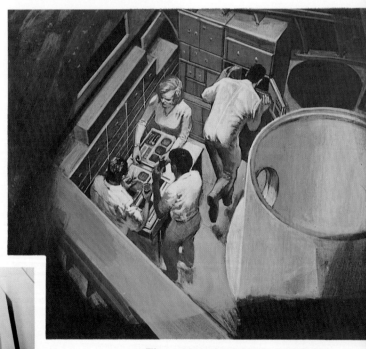

The mid-deck which is just below the flight deck contains several important life support subsystems. Here, a crew of four is having a meal in a rather cozy galley. To the right is clearly visible the airlock which provides access to the cargo bay.

The official name for the Space Shuttle toilet is the Waste Collection System. The zero-gravity environment makes its operation somewhat more complicated than your own bathroom facility. Nevertheless, if you follow the instructions carefully before using it, all should go well.

An Orbiter technician in a spacesuit and backpack with a Manned Maneuvering Unit is replacing an electronics module from an orbiting payload. On-orbit repair and replacement services will be available to meet the needs of new satellites in the 1980's.

With a disabled Orbiter in the background, a rescue crewmember in a spacesuit with backpack transports another crewmember in a rescue ball.

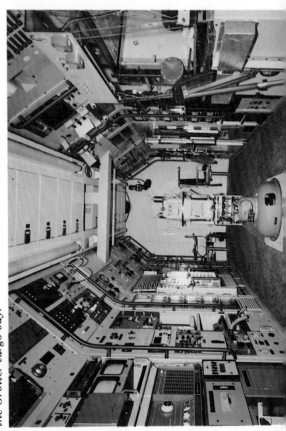

The Orbiter crew prepares to retrieve a large satellite for repairs. The remote manipulator arm is extended and ready to grab the satellite. It will then place the craft in the cargo bay and return it to Earth for repairs and later relaunch.

This mockup shows what the Spacelab module might look like for its flights in the Orbiter cargo bay.

Return to earth is begun with a retro-fire maneuver using either one or both of the Orbital Maneuvering Subsystem engines. After this burn, the Orbiter is turned around so that its nose is facing forward. The actual entry into the atmosphere is associated with aerodynamic heating of the lower surface and nose of the craft.

An Orbiter returning from space lines up for a final approach onto the runway at the Kennedy Space Center in Florida. Launch Complex 39 is plainly visible to the right. The Vehicle Assembly Building is also in view for the crew to see as they come in for a landing.

Imagine yourself in the twenty-first century and the resident of a space colony just returning from a holiday on Earth. The 19-mile long and 4-mile diameter cylinder at the right and its twin at the left are seen as they would appear from an approaching spaceship some 20 miles away. This concept of a space colony orbiting between the Earth and moon was suggested by Professor Gerard K. O'Neill of Princeton University. By their very nature, space colonies will become independent of Earth. They will be self-sustaining and have their own government. Experiences of human nature tell us that we may well someday have real "Star Wars."

CHAPTER FIVE

RETURNING FROM ORBIT

WHAT TIME IS IT WHEN WE GET BACK

One nice thing about a trip into earth orbit, you only need a one-way ticket. At the end of the flight, you should be back home on earth. I have been asked the question many times, "How fast does time pass when in orbit?" People are surprised when I tell them that the time on a spacecraft is the same as the time on earth. In other words, if the Shuttle is up for one day of earth time, then the astronauts onboard are one day older. Some people think that when flying east you gain a day every time you pass the international dateline in the middle of the Pacific Ocean. The truth is that the Orbiter circles the earth every 90 minutes and that is the amount of time that passes in the spacecraft and on earth. So when the Shuttle returns after a one day flight, it will come back on Monday if it left on Sunday. We are about to come back from a typical orbital flight.

TARGETING TO A LANDING SITE

Hopefully, each orbital mission will succeed in carrying out all objectives. The schedule will be worked out such that the Shuttle will be in the proper position for a return to a designated landing site. Thus, on a normal mission, all orbital operations will be completed with the storage of equipment and other gear and, if appropriate, securing of retrieved payloads. The bay doors are closed

and sealed, and the commander orders preparations for retro-fire which is needed to initiate the reentry sequence through a slight reduction in orbital speed. This is all it takes to begin the process of coming back home. Checklists are carefully reviewed to make sure that nothing is overlooked. Since the Space Transportation System became operational in 1982, the Orbiter can land at the Kennedy Space Center and, after 1984, at the Vandenberg Air Force Base. Before that time, the first four test flights used Edwards Air Force Base and the Northrup Strip at White Sands Missile Range. These and other landing sites may be used in cases of aborts or non-availability of primary sites. Others are located at Anderson Air Force Base in Guam and at Hickam Air Force Base in Hawaii. A decision to change the planned landing site must be made before reentry. For example, there may exist unfavorable weather conditions at the primary site, or there may be an emergency onboard that requires immediate de-orbit. The Orbiter is committed to its landing site once it has entered the atmosphere.

A description of the return to earth is best offered with a specific mission as an example. Let's take a look at the events associated with the first orbital test (STS-1) flown in 1981. The primary landing site was Rogers Lake bed at Edwards Air Force Base. All conditions were right and runway 23 was used for the landing. This has an elevation of 2,200 feet above sea level and is located at 34.9 degrees north latitude and 117.8 degrees west longitude. The de-orbit thrust maneuver occurred some 53 hours after liftoff as the Orbiter passed 40 degrees south latitude and 63 degrees east longitude over the Indian Ocean. Yes, this is half-way around the world from the landing site. You don't just drop in from orbit to the landing site. Careful planning and precise maneuvers are required to make a successful return possible. At de-orbit, the Shuttle was in a 40.3 degree inclined orbit with an altitude variation between 171 miles and 177 miles. The precise location and timing of the de-orbit maneuver was selected to provide several minutes of tracking by a ground station at Guam just after retro-fire was completed. This precaution permitted the Orbiter to communicate with the ground to update its reentry parameters at a critical time. The best alternative presented for backup de-orbit would have been one orbit later in sequence. Choosing this orbit would have severely reduced tracking and communications with the Guam ground station after retro-fire. De-orbit from earlier orbits would not have provided adequate tracking and communications at all after the retro-fire burn. In those cases the Orbiter would enter the atmosphere before communicating with any ground station. De-orbit did result in a landing at 10:20 a.m. local time. There is a ground rule that landings should take place before 10:00 a.m. local time because the region around Edwards is notorious for high winds and turbulence after that time each day. However, the landing time was close enough to 10:00 a.m. that winds were not yet a factor.

The de-orbit maneuver can be carried out with either one or both of the Orbital Maneuvering Subsystem engines since there is sufficient propellant in the tanks to do it either way. The entry-weight of the Orbiter for this mission was approximately 183,000 pounds, and the free-fall time from the end of the de-orbit maneuver to the *entry interface* into the atmosphere was almost 20 minutes. For all retro-fire maneuvers, the Orbiter is oriented in a tail-first manner so that the thrusters are pointed properly. After the burn, the vehicle is reoriented nose-first to the proper reentry attitude. At about 400,000 feet (76 miles) the reentry orientation is es-

tablished and maintained by torques from the reaction control system. Aerodynamic control surfaces are cycled before entry to check out the hydraulic system. Nominal conditions at the entry interface are 3,526 miles to the landing site and 17,500 miles per hour velocity.

You might say that the entry trajectory is shaped by committee. Several factors come into play. For example, we want to minimize the effects of reentry on the Thermal Protection subsystem, maximize the flight control performance, and minimize structural loads while maneuvering to compensate for trajectory errors. In some cases, these objectives result in conflicting requirements for the entry profile. For example, shortening the entry range reduces the thermal protection subsystem effectiveness. Increasing the angle of attack by raising the nose somewhat would make the thermal protection subsystem more effective, but the vehicle becomes less stable and its maneuverability is limited. Between 400,000 and 50,000 feet altitude, the Orbiter can move laterally (cross range) up to about 1,265 miles to reach a designated landing site. In general, increasing the angle of attack reduces the cross-range capability. On OFT-1 only a fraction of the cross-range capability will be used. Nevertheless, the entry profile for this flight will be a compromise, as it will for all Shuttle flights, between conflicting requirements.

REENTRY AND RETURN TO THAT HEAVY FEELING

Reentry into the atmosphere is potentially the most dangerous part of the flight. Anyone of a number of undetected flaws or faults could cause total loss of the Orbiter and its crew. For example, if the Thermal Protection subsystem is damaged during launch or if some unforeseen chemical reaction took place to loosen the glue which holds the 34,000 ceramic silica tiles in place, the vehicle could be destroyed in a fiery ball many miles above the Pacific Ocean. As it is, the adhesive used to bond these tiles to the Orbiter cannot withstand temperatures below a minus 160 degrees Fahrenheit and the dark side of the Shuttle can become this cold while it is in space. This means that the Orbiter cannot stay in one orientation for more than three or four hours at a time. Thus, it must rotate periodically to "toast" these tiles. If the reaction control system does not maintain the correct angle of attack during reentry, overheating of the tile adhesive can likewise break the tile bonds. The commander will have to stay extremely alert during this phase of flight. If there is a failure in the autopilot he will have to take over manually.

Actual entry into the upper bounds of the earth's atmosphere occurs 19 minutes 39 seconds after completion of the retro-fire burn for a nominal case. The reentry profile on STS-1 was shaped so that the temperature limit on the thermal protection tiles will not exceed 2,800 degrees Fahrenheit. Well, you may not raise an eyebrow over 2,800 degrees, but this is much hotter than your fireplace gets in a roaring blaze. This is hotter than your acetylene torch gets. In fact, the skin of the Orbiter is aluminum. The melting temperature of pure aluminum is 1,220 degrees Fahrenheit. So those thermal protection tiles are all that is separating the crew from a fiery end. There is a wide variation of temperature over the lower surface of the vehicle. However, the hottest spot is just below the nose. As the entry proceeds, the crew looks through the viewing ports to see a red glow of air around the forward part of the vehicle. This is when you decide whether or not you are ready for another flight into space.

The Thermal Protection Subsystem consists of over 34,000 ceramic silica tiles. They look something like your bathroom tiles in size, but are very light in weight. These are applied externally, all over areas which will be aerodynamically heated, either during ascent or during reentry. Vertical tail surfaces actually get hot during ascent, up to 2,200 degrees Fahrenheit. The more critical problem is reentry, however, when nose temperatures can go to 2,800 degrees Fahrenheit. After each flight these tiles are inspected and replaced as needed. They are glued into place with a special adhesive, using a unique process to insure that they are secure. Their function is to prevent the heat generated by air friction from getting to the main structure and aluminum skin of the Orbiter.

During STS-1, the crew noticed that several thermal insulation tiles were missing from the OMS pods. These had apparently been ripped away or damaged during launch. Fortunately, the pod areas are not exposed to extreme aerodynamic heating during reentry, and *Columbia*'s return was successful.

The sensation of reentry is somewhat dif-

ferent from that of reaching orbit. In this case the crew starts the entry in a weightless state. As the atmosphere gets denser and denser, drag increases, and a downward force is increasingly felt. Arms and legs begin to feel heavy. The crewmembers are pulled down into their seats. At the peak of deceleration this pull actually exceeds twice their normal weight on the ground. As the speed slows and the Orbiter begins to respond like an airplane, this force reduces to just their weight.

The entry profile is shaped to bring the Orbiter to the terminal descent area with proper speed, altitude, position, and orientation. Deflections of aerodynamic control surfaces are programmed to minimize attitude control torques required of the reaction control system. At an altitude of 250,000 feet, the atmospheric density is sufficient for some aerodynamic control using the body flap, elevons, rudder, and speed brake. This helps the reaction control system which is not deactivated until an altitude of 80,000 feet is reached. Beyond this point the Orbiter is a powerless free-flying glider. The use of aerodynamic control surfaces above 80,000 feet helps conserve propellant in the reaction control system during the entry phase. At speeds above 8,180 miles per hour, the elevons are deflected two degrees up, and the body flap is deflected five degrees up. The speed brake is deflected to a full-out position at a

Many people became very concerned during STS-1 when several tiles were noted to be missing from the OMS pods. If tiles were also missing below the nose of Columbia, where observation was not possible, then reentry could have been fatal. Fortunately, all critical tiles were in place.

Columbia approaches Rogers Dry Lake Runway 23 to end its historic first flight.

THERMAL PROTECTION SYSTEM (TPS)
MATERIALS CONFIGURATION

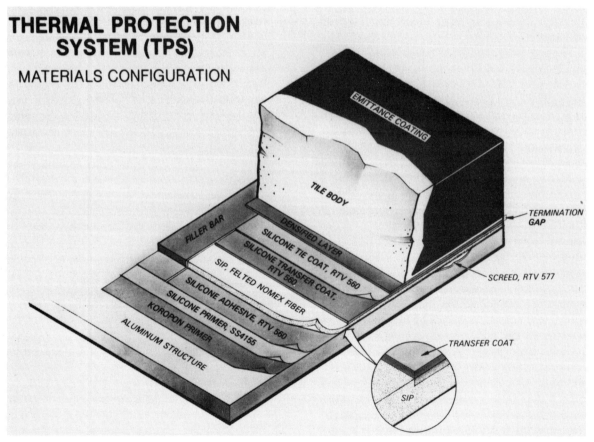

The Thermal Protection System with some 34,000 tiles all over the Orbiter, provides heat protection during ascent and reentry. Even though each tile appears to be a simple block of material, it is, in fact, part of a very complicated lamination of varying materials.

speed of 5,450 miles per hour to induce a nose-up torque so that the elevons can be deflected down in this speed range. Conversely, at a speed of 2,045 miles per hour, when the elevons are moved to an up position, the speed brake is moved to a smaller deflection to reduce the nose-up tendency. At subsonic speeds, the nominal speed brake deflection is at the mid-value to allow modulation to effect speed control. The body flap deflection schedule is used as a means for trimming the Orbiter attitude.

COMMUNICATIONS AND NAVIGATION

In the last chapter, we saw how the Shuttle communicates with the Mission Control Center while carrying out its tasks in space. Once the reentry sequence is begun with the retro-fire burn, the mode of communications changes entirely. Direct Orbiter-to-ground links are established as the vehicle passes within the viewing area of each station involved. Although this kind of coverage is poor, at best, near the beginning of the reentry phase, only short intervals of time are required to check the entry flight path and status of the vehicle.

The precise instant of retro-fire initiation is arrived at by carefully matching the reentry profile and the site location with orbital properties. Once the Orbiter entry begins, the stage is set for the entire return to Earth. Orbiter computers keep track of position, speed, altitude, and orientation, ever so smartly, to bring this bird through the upper atmosphere. Communications and navigation are important in order to reach the landing site safely. As the Orbiter makes the transition from a reentering spacecraft to a powerless aircraft, it begins to use conventional navigation facilities. In fact, it uses the very same navigational aids as do all airplanes, large and small, in the United States. These facilities, commonly called VORs or VORTACs by pilots, are located around the country and provide direction, distance, and speed information continuously. Only a few of these stations will be necessary for any particular flight.

To better understand how this works, let's review the communications and navigation plan for STS-1. The landing site was Edwards Air Force Base, and the runway was 23. Because of orbital geometry, reentry is from the west, over the Pacific Ocean and the West Coast, just north of Los Angeles. Approximately six minutes after completion of the de-orbit burn, a ground station at Guam is able to pick up signals from the Orbiter as it comes up over the horizon. However, this link lasts only six minutes because the Shuttle races across the sky to the other horizon. Another seven minutes later, the entry point is reached. Three and one-half minutes after reaching this point, a communications blackout is encountered at an altitude of about 50 miles and a ground speed of about 16,700 miles per hour.

Communication *blackouts* have occurred routinely on all previous reentries. They are caused by the extreme heat generated around the Orbiter. Air molecules become so excited by this heating that they lose charged particles, that is, they become *ionized*. This results in the formation of a layer of positively and negatively charged particles around the vehicle. Such a layer is called a *plasma,* and it acts like an excellent shield against radio signals, both incoming and outgoing. As long as this heat is generated so intensely, the plasma will persist. The blackout lasts for about ten minutes and ends several hundred miles out over the Pacific at an altitude of about 34 miles. The Orbiter's speed at this point is approximately 8,200 miles per hour. Shortly thereafter, radar stations at Pt. Pillar near Half Moon Bay, California, and at the Vandenberg Air Force Base picks up the vehicle's signal. From this moment on, the Orbiter is tracked from the ground, and its position and velocity will be known continuously. At about the same time, a communications link is established with the Buckhorn Station located at Edwards Air Force Base. As soon as voice contact is made, latest runway information is forwarded to the crew. Wind shifts and other conditions may cause a last minute change in landing runway selection. If the runway is changed, the crew will have to begin maneuvering at an altitude of 30 miles and a speed of 6,485 miles per hour.

Radar is used to track and guide the Orbiter until the standard navigation aids can be used. About 200 miles west of the coast these devices begin to be received by the Shuttle crew. Signals from the navigation radio (VORTAC) at San Luis Obispo are received first. These give distance, speed, and direction information to the aircraft-like autopilot in the Orbiter. Then the Fellows station (southwest of Bakersfield) is received. Finally, the station right at Edwards Air Force Base is picked up and used to "home" into the field. A total of ten such stations are used for navigational information to the landing site over the last 250 miles.

THE HOME STRETCH

The first return flight of a Shuttle from orbit landed at Edwards Air Force Base at 10:20 a.m. on April 14, 1981. It is doubtful that any passerby noticed the Orbiter as it crossed the coast just north of San Luis Obispo, California because it is still 26 miles high. Many spectators try to spot it, but this is extremely difficult while it is at such a high altitude. After passing the coast, the vehicle continues inland over the Sierra Madre Mountains, past Fellows, California, and on over the desert south of Bakersfield. It is still over 90,000 feet high at this point, but descending rapidly. Onward it travels, over the Tehachapi Mountains and just south of Mojave, California. It continues southeast and overflies Edwards descending past 54,000 feet. This is followed by a left turn to the southwest and onto a final approach course with runway 23. The Orbiter is not visible to the naked eye until over the Tehachapis as it descends below 50,000 feet. While making the left turn, it passes through the 25,000 feet altitude point. At this point, everyone sees it clearly. The Orbiter is returning from space for the first time. In just two minutes its wheels should touch down. The welcoming crowd awaits silently, breathlessly, as the vehicle initiates the preliminary flare while still at 2,000 feet above the runway. Landing gears come down at 300 feet followed by final flare at 50 feet, and touchdown occurs. A loud cheer can be heard around the world.

The *approach and landing phase* is that last part of the descent when aligned with the runway and below 20,000 feet altitude. A 22-degree glide slope is used in the steep descent phase. In fact, the vertical speed is in excess of 140 miles per hour during this phase. A skydiver freefalls at about 120 miles per hour. Thus, a freefalling skydiver would be passed right by as the Orbiter makes its way to runway 23. A pre-flare maneuver is executed at about 2,000 feet to bring the glide slope to a shallow 1.5 degree angle. This is followed by a final flare maneuver just before touchdown. During steep descent, a speed of 340 miles per hour is maintained so that the pre-flare phase will last about ten seconds before final flare.

All approach and landing maneuvers can be done by the autopilot or manually by the pilot. However, in most cases the pilot will be monitoring the autopilot and will be able to take over if required.

The touchdown conditions for a normal landing include a speed of 220 miles per hour with a sink-rate of three feet per second (180 feet per minute). Wheels should contact the pavement 2,000 to 3,000 feet from the runway threshold. This ideal touchdown speed is arrived at by making allowances for atmospheric conditions at Edwards and it includes temperature variations and turbulence. In a "worst-case" situation, the Orbiter would make it to the threshold at a speed of 193 miles per hour. Even this is a relatively high landing speed. For example, U.S. Air Force jets land at speeds between 155 and 178 miles per hour. A DC-9 lands at 129 miles per hour. The final sink-rate of three feet per second was chosen to insure a firm touchdown and to eliminate any floating due to ground effects. Since drag increases when the landing gear is extended, the gear is not lowered until the latest possible moment. So not until an altitude of 300 feet is reached is the gear switch thrown, slowly rotating the wheels into position. Just seconds before touchdown the three struts lock into position. Spectators heard "squeak, squeak, squeak" as the tires contacted the ground, and the Orbiter was home.

The first two orbital flights ended with landings at Edwards. STS-3 landed at Northrup, and STS-4 returned to Edwards. Beginning with STS-5 landings may be made at the Kennedy Space Center.

Similar procedures will be used for communications and navigation in future flights, but different ground stations will be involved because of the geographic differences between the landing sites. Approaches will again be from the west with the trajectory passing near Tampa, Lakeland, and Disney World in Florida. By the mid-1980s, Vandenberg will be operational, and Shuttles will be arriving and departing regularly. Because of the southerly departure directions from Vandenberg, Orbiters will be able to approach for landing from the south. One might ask, at this point, why Kennedy launches to the east result in approaches from the west, and southerly launches from Vandenberg result in approaches from the south. The answer is that we will take advantage of the earth's rotation

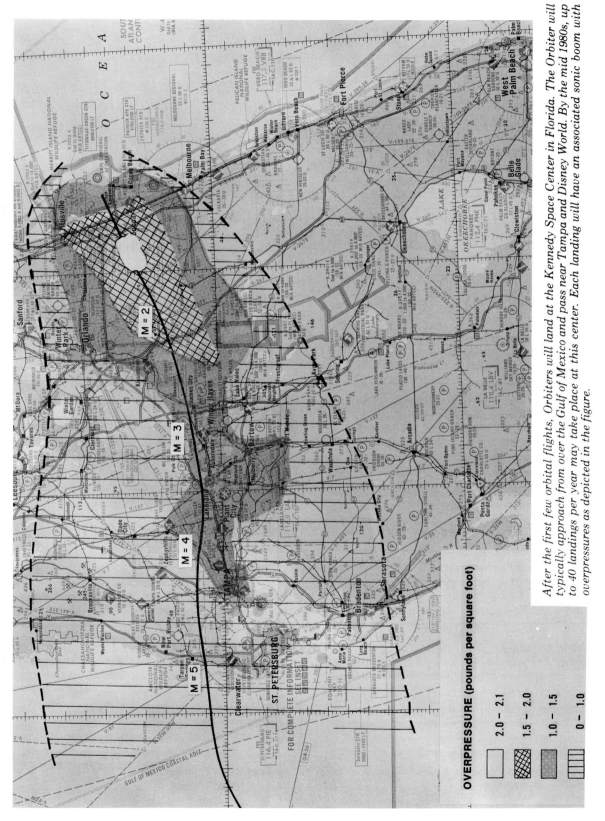

After the first few orbital flights, Orbiters will land at the Kennedy Space Center in Florida. The Orbiter will typically approach from over the Gulf of Mexico and pass near Tampa and Disney World. By the mid 1980s, up to 40 landings per year may take place at this center. Each landing will have an associated sonic boom with overpressures as depicted in the figure.

OVERPRESSURE (pounds per square foot)

☐	2.0 – 2.1
▨	1.5 – 2.0
▒	1.0 – 1.5
▤	0 – 1.0

in those cases where we have high inclination orbits. In other words, by letting the earth pass under the orbit and using precise timing, the Orbiter can approach Vandenberg either from the north or from the south. However, it is not desirable to have approaches from the north because the reentry path would take the vehicle over Russia on its way down. Soviet tracking stations would very likely pick up the Orbiter as an intercontinental ballistic missile on their radar screens. The thought of this possibility and potential Russian reaction makes our State Department very nervous. After all, we wouldn't want the Russians to launch a retaliatory attack on the United States just because we're bringing back one of our Shuttles. Of course, there are other reasons related to communications, tracking, and environmental effects for eliminating approaches from the north.

In all cases, overpressures from a normal return trajectory will not exceed 0.5 pounds per square foot until the Orbiter is within 575 miles of the landing site. Overpressures of one pound per square foot are exceeded at about 104 miles out. No orbiter return overpressures should exceed 2.1 pounds per square foot, and the area associated with overpressures in excess of two pounds per square foot is limited to about 100 square miles within 28 miles of the landing site. The total area affected by the reentry boom is about 7,000 square miles. In Florida, an approaching flight from the west would expose about 500,000 people to a bang which would be weak for most who hear it. Each of the orbital flight test landings at Edwards exposed approximately 50,000 people since most of the boom impacts the sparsely populated region northwest of Los Angeles. Landings at Vandenberg affect only a few thousand people because the boom is primarily over the Pacific Ocean.

Many people will wonder how NASA can get away with producing these sonic booms while there is a U.S. code of federal regulations which prohibits civil aircraft, including government airplanes which carry commercial cargo, from creating sonic booms over the United States. Is the Space Transportation System subject to this regulation or is it exempted by the Federal Aviation Administration? In March 1977, the Chief Council of the FAA, in a letter to NASA, concluded that the Orbiter is not an aircraft within the meaning of the FAA Act of 1958. Thus, it is not subject to the sonic boom prohibition.

This was certainly a fortunate development for NASA but possibly not so fortunate for many people who live in Florida and California. In any case, it would be difficult, if not impossible, to eliminate these sonic booms and still have a Space Transportation System with an economically acceptable performance capability.

COMING BACK WITH A BANG

The environmental effects of reentry and descent through the atmosphere are minimal. Only the sonic booms produced by the Orbiter are of concern. The first four landings were at desert locations, with booms impacting areas northwest of the fields. Intensities of up to 2.1 pounds per square foot were experienced within ten miles of the landing site. As it was pointed out in Chapter 3, this level causes a mixed pattern of startling and eye blinking effects in about one-half the people exposed. No structural damage to buildings is expected, but loose windows may rattle for a few seconds.

Starting by 1983, landings will be at the Kennedy Space Center, six per year at first, then up to 40 per year by the mid-1980s. Each landing will have an associated sonic boom. Orbiters which were launched from the Cape will produce a peak overpressure region a few miles northwest of Melbourne, Florida. Again, up to about 2.1 pounds per square foot will be experienced. Orlando and Disney World will experience overpressures of up to one pound per square foot. Since the approach path will take the Orbiters over Tampa and Lakeland, Florida, there will be a wide footprint of exposure of up to one pound per square foot all across the Florida peninsula. If you happen to live in this region or are visiting Disney World, a returning Orbiter definitely announces its arrival with a loud bang. If you then look up to see it, remember that the sonic boom reaches you after the Shuttle passes. So look to the east and use binoculars if you have them since it will still be several miles high. Occasionally, an Orbiter will be launched from Vandenberg and will land at Kennedy. Such flights will probably be brought in from the south. This will produce a loud bang of varying intensity from Miami to Melbourne. The reverse flight plan will also take place once in a while, a flight from Kennedy to Vandenberg. These would bring the Orbiter in from the Pacific with essentially no boom impinging on land.

POSTFLIGHT OPERATIONS

Once the wheels contacted runway 23, brakes were applied and *Columbia* stopped some 9,000 feet later. In a typical mission *post-flight operations* begin immediately in order to render all systems safe and to start the recycle process for the next flight. Assuming that all has gone well and the Shuttle arrives safely at Kennedy or Vandenberg, the Orbiter is taken to the processing facility. Fluid and propellant are drained from the various systems, and all plumbing and tanks are cleaned. Any special equipment that was needed for the last flight is removed or checked if to remain onboard. Returning payloads are removed and delivered to the user, prepared for the next flight, or stored.

In the event of an Orbiter landing at a secondary site or at a contingency site, the responsible base will dispatch equipment and transportation to remove payloads that will return from orbit. A contingency landing site will not have any special equipment available except for that necessary for crew survival and Orbiter towing. The 747 carrier aircraft will also be dispatched to bring the Orbiter back to its primary site.

From this point on, all activities are those described in Chapter 2. We have now completed the cycle of a Shuttle mission from the pre-launch activities on through the complete flight and back to the landing site. At this point you might ask: "Where has all this space flight gotten us?" The last two decades have reaped uncountable benefits for us on earth. The next chapter delves into only a few that affect us in our daily life and will have a significant influence on our future.

Home-sweet-home for most Shuttle vehicles is the Kennedy Space Center. Returning Orbiters will land on the 15,000-foot runway in the background. Preparation and processing facilities are shown in the foreground. Launch Complex 39 is to the right of this area.

CHAPTER SIX

FOR ALL MANKIND

INVESTMENT IN AMERICA'S FUTURE

Neil Armstrong said it so aptly, ". . . for all mankind." This is what the space program is really all about. How quickly we forget! If you recall watching the evening news on television after a major space mission, then you will remember that the networks like to ask the "man-in-the-street" question about the value of the space program. All of us have heard the response, "the space program is taking money that could be better spent for more pressing programs such as social welfare needs." Another comment is, "Why are we sending all of this money into space?" It is indeed difficult for a person not directly involved in the space program to defend it spontaneously. All the benefits that we have received are not labelled "space benefits." Dr. Wernher von Braun summarized the situation very aptly in a statement to the U.S. Congress: "Direct benefits from space technology already have entered our lives to a far greater extent than most people realize. Advances in medicine, communications, earth resouces, environmental monitoring, and thousands of new products have given the United States technological leadership in the world today."

Let's go back and take a look at the man-in-the-street response. What about all that money that's spent on space programs? Many people believe that, if we took this money and put it into our social and welfare programs, it

would solve all of our domestic problems. However, the facts do not confirm this. For example, in 1969 the U.S. Government spent $65.2 billion on social action programs and $4.2 billion on space. In 1972 the federal budget called for social action spending in the amount of $100 billion while the space budget fell to a low of $3.2 billion. In 1978 the budget called for $160 billion to be spent on social security and welfare, while only $3 billion went to the space program. Alas, this trend continues, even though the NASA space budget is now over $5 billion. Fifty times as much money goes into social welfare programs than into the space program. Obviously, increasing the social action budget by less than two percent is not going to solve all of our economic and domestic problems. Of course, there are still those who say that this little amount of money would help. Yes, it would help. However, I further contend that the space program is not even costing us a relatively small amount of money, but it is actually paying for itself many times over. This will be backed up very soon with examples of the benefits to our everyday quality of life.

Are all those dollars going into space? In case you haven't noticed, nobody will accept American dollars in space. There are no stores, gift shops, or tourist "traps" in orbit. There isn't even a general store on the moon.

So where are all those dollars going? They are going into over 20,000 companies all around the 50 states. This has resulted in as many as 300,000 jobs at the peak of activities in the 1960s. Contracts have gone to all states in our nation. Every dollar allocated to space has been spent right here on earth.

Some of us will wonder how people can object to investing a small part of our Gross National Product in technology. Opposition to this kind of spending is not new. For example, an excerpt from a commission report on the evaluation of a proposal states the following: "The committee judged the promises and offers of this mission to be impossible, vain, and worthy of rejection: that it was not proper to favor an affair that rested on such weak foundations and which appeared uncertain and impossible to any educated person, however little learning he might have." This is an excerpt from the report of the Talavera Commission in Spain in 1491 which considered a proposal by a fellow named Columbus. Fortunately for all of us, Isabella had a bit more foresight. She showed us one of the messages of history which is eminently relevant to the United States today. De-emphasis of technology was associated with the decline of the Roman Empire and, more recently, other European countries. Few undertakings in the modern world are so important as the development of technology. It is not technology in itself that makes a nation great. It is its benefits, its spinoffs, and its influence around the world. Technology, like that generated in the space program, is now a currency of foreign affairs. It is a tool of advanced nations and a hope for under-developed ones.

A free society needs pace setters in multiple activities. The alternative is mediocrity. No other technological endeavor has set as high standards as are required in the space program. The term "zero defects" is an invention of space technology in which machines must function perfectly in the almost impossible environments of vacuum and temperature extremes. Space has its own purposes, targets, and destiny. This technology has already accomplished things that cannot otherwise be done economically or, perhaps, done at all. However, it is not only those spectacular events that we see on television that make it all worthwhile. Even more remarkable is the often overlooked fact that the pursuit of space goals has generated innovations in virtually all fields of science and technology. It has helped stimulate progress in areas not even remotely connected to the original program. In the following sections we are going to see just a few examples of how our lives have been affected on a daily basis by this small investment in space technology.

THE SPACE PROGRAM AND YOUR JOB

The readers of this book include those of you who have a variety of jobs and professions not obviously related to our space activities. If you are one of these people, it is highly likely that you have not thought of ways in which your life or your job is affected. The relationship between building a space capability and improving life on earth is closer than you might generally realize. We all know that the most critical problem areas in U.S. society today are inflation and unemployment. This is a nation of workers, despite our relatively high rate of unemployment. Even at that, there are over 90 million Americans now employed. It is extremely difficult to legislate and to administer the creation of new jobs in a free enterprise system. There is a delicate interplay of capital investment, technology, and talent. How effective is the space program in creating and improving jobs while reducing the rate of inflation? Let's find out.

Economists are somewhat like weather forecasters. No matter how many you ask about tomorrow's weather, each will give you a different opinion. There is one thing that economists agree on, however, *productivity* is a potential long-term solution to inflation. The forces creating inflation are neutralized by getting more out of each unit of labor and by expanding the supply of goods and services. If space funding were diverted into social welfare programs, would inflation be decreased? The answer is no. A couple of years ago a comprehensive study showed that just the opposite would be true. An increase in funds for space research and development would, in fact, reduce inflation. In general, high technology endeavors, of which the space program is clearly the largest, result in significant rates of social return far out of proportion to their cost. For example, a one billion dollar annual increase in space research and development would have the following effects in ten years. The Gross National Product would increase by at least $23 billion, equivalent to an annual return on our investment of over 40 percent. The inflation rate would fall by two percent. Unemployment would be reduced by 400,000 jobs.

This appears to be a tremendous payoff for a small investment. The reason these payoffs are so high is that the research investment spreads to other industries as well. Achieving higher industrial output and lower inflation is inevitable due to the growth of labor productivity. This means that less labor is needed per unit of output. As less labor is required, the costs decline, leading to price decreases and increases in consumers' real income. Thus, greater purchases of goods and services result. The chain reaction eventually leads to improved mass production and further lowering of the unit costs. Moreover, the size of the labor force can increase through greater job opportunities spread across many industries, old and new. Unfortunately, such economic spinoffs don't occur immediately. It takes about five years for the effects to become evident. For this reason in particular, our confidence in space program investments must be of a continuing nature. The fact is that our growth and output per labor hour has fallen behind that of other industrial nations due to a slowdown in the past 25 years of U.S. investment in new technology. We must revitalize our space investment as part of the larger emphasis on industrial growth if the U.S. is to remain the technological leader in the world.

At the peak of development activity, there were 50,000 people in 47 states working on the Space Transportation System. Nevertheless, jobs directly related to an on-going space program represent only a small part of the job-making and inflation-reducing payoff from space research. Spending on space must be distinguished from investing in space. We don't just want to make jobs but to invest in making whole new industries. The computer business is probably the greatest non-aerospace spinoff of space technology. Every major computer system in the world is made in America. A million large computers were operating by 1980. This reflects the growth rate of 40 percent a year since 1976. This growth in computer development and use magnifies the increased productivity throughout the world in industries using computers.

Of course, many industries have been affected by the space program. The high standards required of orbital missions have advanced the state of vacuum technology so far that processing of high-purity materials, coatings, and semi-conductors is now routinely done in vacuum. The *cryogenics,* or ultralow-temperature, industry has erupted as a direct result of liquid gases needed for rocket propulsion and life support in space. Hospitals and steel mills are among dozens of beneficiaries throughout the nation that now use liquid oxygen, nitrogen, and helium on a routine basis. Even the pipeline business has benefited from new aerospace technology. A method of winding glass thread to make light-weight rocket casings has been applied to the manufacture of lightweight sewer pipes, making it possible to lay entire pipelines by helicopter, especially useful in remote areas. I happened to be working with one such company when they started developing these glasswound pipes. Personnel there jokingly claimed that their new motto was "our business is going down the tubes."

A few years ago, economists at Mathematica Inc. isolated four examples of spinoffs and estimated their return to the economy. Benefits from these were projected conservatively to add up to $7 billion by the early 1980s. These are actual thriving industries which were spun off from the space program. The biggest growth industry is *integrated circuits*, a technology developed for satellite controls, communications, and other space uses. These devices are now used in television sets, automobiles, and hundreds of industrial and household products. The return from improved technology is estimated at over $5 billion between 1963 and 1982. Another industry is that of *gas turbines*. These are now widely used for electric power generation and have saved an estimated $111 million between 1969 and 1982. A *structural-analysis computer program*, originally developed to help design more efficient spacecraft, is being used today to design railroad tracks and cars, automobiles, bridges, skyscrapers, and many other structures. This program is expected to return over $700 million in cost savings between 1971 and 1984. The last example is *insulation for cryogenic uses*, mentioned in the preceding paragraph. The probable benefits are in excess of one billion dollars between 1960 and 1983. Remember that these represent only four examples of the many spinoffs associated with the space program. We are now about to get more specific about space benefits. Examples of effects on your daily life are presented in the following sections.

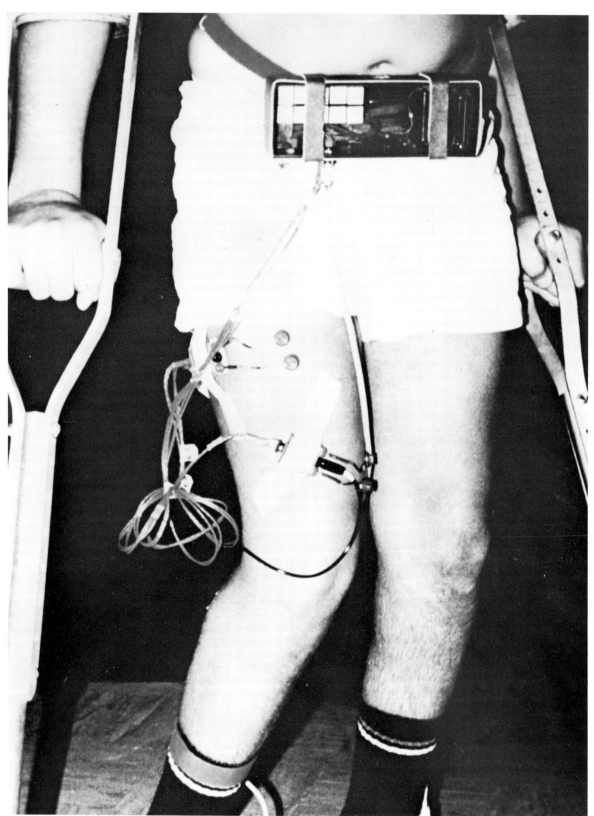

Space telemetry has been used to develop a complete motion analysis laboratory to analyze the walking patterns of crippled children. Biotelemetry has been adapted to monitor the awkward, jerky gait of cerebral-palsied children, thereby eliminating the previously necessary long bundle of wire leading to the recording equipment. In order for corrective therapy to be effective, precise knowledge of how each muscle group contributes to the child's walking problem is required.

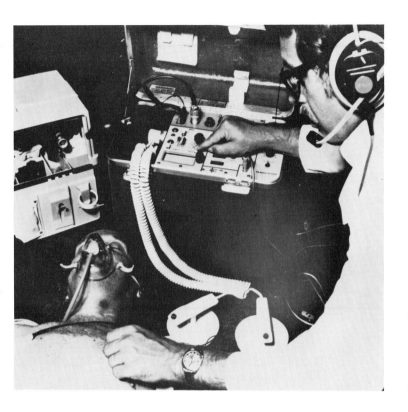

A checklist devised for the Skylab inflight medical support system has been transferred for use in public emergencies. Paramedics assigned to the Houston fire department now are using the checklist.

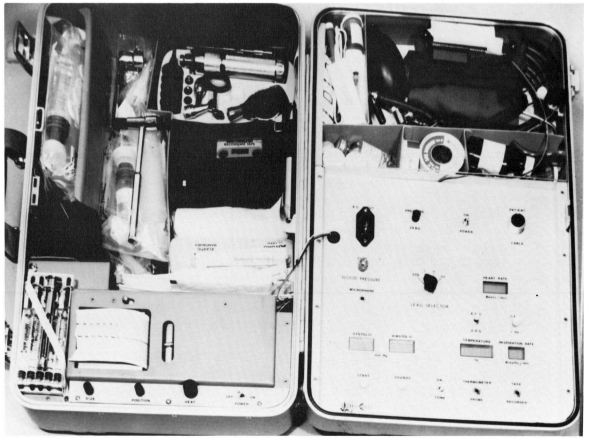

The portable medical-status system contains an electronic vital signs monitor, a machine for recording electrocardiograms and electroencephalograms, equipment for minor surgery, and conventional diagnostic instruments such as a stethoscope. The unit weighs 30 pounds and measures 7x22x14 inches.

YOUR HEALTH

Space medicine, together with innovations in remote acquisition, monitoring, and interpretation of physiological processes during manned flights, has generated technology for improving both the quality and quantity of health care here on Earth. Even in this advanced country of ours, the problem is staggering. Seventy percent of all Americans visit their doctors at least once a year. One out of ten is hospitalized each year. The cost of this country's health services is over $80 billion annually, about eight percent of the Gross National Product.

Monitoring techniques similar to those used for the Apollo astronauts are routinely used in hospitals today. This radio-linked automatic patient monitoring system can collect several types of physiological data from as many as 64 patients simultaneously. This information can be relayed to a central computer and almost instantly be processed for use by the medical staff.

We have all heard of pacemakers and know that they control the heart beat for many people who have irregularities in their cardiac system. Modern pacemakers are actually spinoffs of our space program; they are an outgrowth of miniaturized solid-state circuitry developed for spacecraft. When the natural heart beat becomes irregular because of heart disease, the electronic pacemaker delivers small, regular electric shocks to pace the heart. About 30,000 pacemakers are implanted each year in the United States and a like amount around the rest of the world. Until recently, pacemakers lasted only about 22 months, after which their battery power was depleted. In the past this would require surgery to remove and to replace the device. Repeated surgery can be traumatic and, of course, expensive. Each of these operations could cost more than $2,000.

A new pacemaker, developed by industrial researchers who were assisted by NASA, is rechargeable through the skin by inductance. Once a week the patient simply puts on a charging device for an hour to recharge the pacemaker battery. Since recharging can be done frequently, only a single cell is required in the device, thus reducing its size by a factor of two. The whole pacemaker now weighs only two ounces.

These new pacemakers represent a significant improvement in health care, but the more relevant problem is cardiovascular disease. This is the number one killer in the U.S., accounting for more than a million deaths a year. Many of these could be prevented if firemen and other rescue teams could be better trained and have access to the latest technological devices. A spinoff of Skylab telemetry has led to "Telecare," an emergency system which contains all the instruments that a doctor or paramedic could reasonably want in a cardiopulmonary emergency in one package. The kit contains a respiratory resuscitation system, a fifteen-minute oxygen supply, an electrocardiogram display and telemetry system, a defibrillator for external heart stimulation, and blood pressure measuring system similar to that used for Skylab. Telecare units are now being used in ambulances in Houston, Cleveland and several other cities. Perhaps you will recall the television program, "Emergency." This series depicted paramedics using equipment like that in the telecare units to revive and treat a variety of heart attack and accident victims in the Los Angeles area.

Another cardiac spinoff has been derived from biomedical electrodes used with the astronauts. Over the many years of manned spaceflight, more comfortable and effective electrodes have been developed. These are used today in electrocardiograms and other hospital instrumentation throughout the country. NASA's development of spacecraft *transducers*, devices used to convert energy from one form to another, now have been transferred to detect *arteriosclerosis*, or hardening of the arteries. In the past, the test for this involved inserting a hollow needle into an artery and directly measuring the arterial pulse, a time-consuming and painful procedure. The space spinoff electrodes allow determination of arterial flexibility by external means. The device uses a pressure sensitive translator that converts arterial pulses into electrical signals. When amplified they can be recorded on a standard electrocardiograph.

In recent years the public has become more aware that excessive taking of x-rays for medical examinations can be dangerous. Fortunately, space technology is helping to reduce the quantity of x-rays which a person may be exposed to during such examinations. Until recently, doctors have typically taken more than one picture in order to get proper

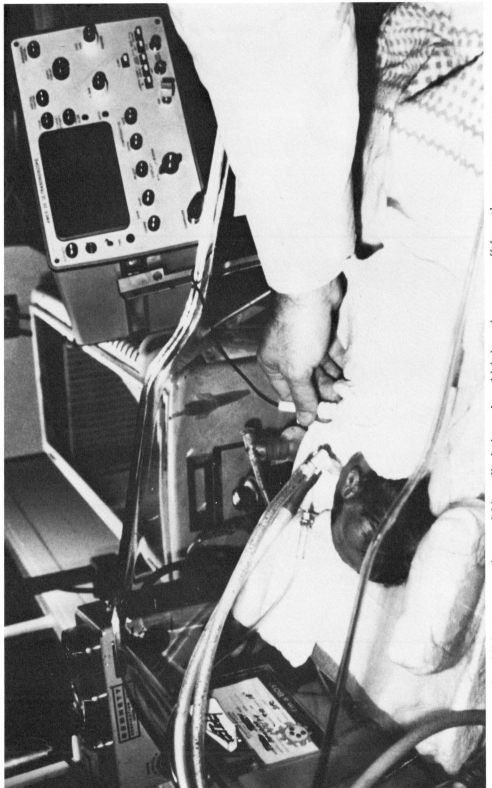

In addition to pacemakers and biomedical electrodes which have been spun-off from the space program, an echo-cardioscope has also emerged. NASA engineers developed the instruments to monitor cardiac functions of astronauts in flight. It forms images of internal structures, using high frequency sound in much the same way that submarines detect underwater objects with sonar. The old method used plastic tubes which were inserted into blood vessels near the heart and involved injecting dye before x-rays could be taken. Babies are particularly susceptible to electric shock hazards and repeated doses of x-rays. The battery powered ultra-sonic device has inherent safety advantages for both adults and children.

A spin-off of the astronaut's biological isolation garment now protects hospital patients who are vulnerable to infection for up to several hours while away from their sterile habitats. A germ-free environment is carried with them. The air is supplied through a flexible tube after it is purified by a system using rechargeable batteries.

exposure. Now, solar cells that convert sunlight to electricity on satellites are being used to help in making a single x-ray exposure sufficient for the medical examination. Since solar cells are sensitive to x-rays as well as light, a sensor has been made from such a cell and placed directly beneath the x-ray film. This determines exactly when the film is properly exposed. Not only has the x-ray hazard been reduced significantly in a trial project at a Pasadena, California hospital, but the number of patient examinations has doubled. These sensors are especially useful in breast radiography. Since the breast is transparent to x-rays, very low-energy rays must be used, requiring precise exposures.

To further reduce the exposure to x-rays, space technology has brought us something even better than the solar cell sensor. A portable "echo-cardioscope" has been adapted for use on Earth after being developed for monitoring heart functions of astronauts. It replaces the need for inserting tubes in the blood vessels in connection with taking x-ray pictures. The device forms images of internal organs using high frequency sound in somewhat the same way that underwater objects are detected using *sonar*. Other ultrasonic imaging equipment is capable of resolving body tissue images that are comparable to the quality of those from x-rays. The first clinical applications dealt with the detection of breast cancer.

You may recall that during the early Apollo lunar flights the astronauts wore special suits between splashdown and their quarantine on the recovery ship. These were designed to protect against unknown micro-organisms. Similar suits now protect immune-deficient children and patients suffering from leukemia, burns, or other illnesses where infection can kill. The suits make it possible for such patients to leave their isolation rooms for several hours. Yes, there are medical benefits for the healthy and the ill, and there are benefits for the young and the old. If we let it, the space program will make us all healthier.

YOUR MOBILITY

To a large extent the National Aeronautics and Space Administration is in the transportation business. It engages not only in delivery of scientific payloads to earth orbit and to the planets, providing services to a variety of users on the Space Transportation System, but also in research related to air transportation.

Much of its work in overcoming technological difficulties is applicable to highway and railway traffic. For example, the Apollo program was very active in analyzing trajectories and landing locations on the moon. This work has contributed to the nation's first fully computerized automobile traffic control system. A prototype system has been installed by an aerospace company in a nine-square-mile area of Los Angeles County. There it controls about 200,000 vehicles daily at 112 intersections. It operates by gathering information on existing traffic conditions. Data about conditions are fed into a computer which calculates the best traffic light sequence to match the traffic flow. Motorists creeping ahead in rush hour traffic find that the system relieves congestion far better than the usual predetermined red-green light pattern. Not only does this increase the average speed of traffic and relieve anxiety, but tests have shown that a considerable savings is realized in reduced fuel consumption and auto maintenance due to engine idling. An added benefit is that the amount of auto exhaust pollution is likewise reduced. Similar systems are now being installed in Baltimore and other cities around the country.

When the two Viking landers parachuted to the surface of Mars, each one-ton vehicle was supported by only three straps of a remarkable new fiber. This material is five times stronger than steel and is now being used to make the cords of new winter tires for automobiles. The rubber in these tires is the same as that used on the wheels of the "Rickshaw" which accompanied Astronauts Alan Shepard and Stuart Roosa on their lunar trip in order to transport equipment across the surface of the moon. Conventional tires lose their flexible qualities below freezing temperatures. These new ones provide traction even in the coldest weather. They remained pliable on the moon in temperatures as low as 195 degrees below zero.

Smoother rides on highways and railways may soon be reality through the adaptation of gyrostabilized, spacecraft guidance systems. A new instrument is currently being developed to measure bumps, curves, and grades. Highways and railways may soon be shaped more precisely for better and safer travelling. Such improvements will also make it easier to maintain our roads and rails. Other automobile-related improvements from space include better brake linings, consisting of a

Technology from a spacecraft malfunction detection system has provided the basis for a commercial product which diagnoses vehicle engine problems. It is called "Autosense," and it examines the vehicle's engine and provides a computer printout comparing each component with factory specifications. Thus, correct repairs are indicated. This same instrument verifies that problems have been corrected.

NASA research has led to the creation of the brightest hand-held spotlight ever developed, called the Stream Lite-1 Million. The unique xenon spotlight has a peak capacity of one million candle power, about 50 times brighter than a high beam light of an automobile. The 7-pound device can operate on a 12 volt battery. It is especially useful in penetrating fog and smoke.

NASA has used space technology to develop a driverless electric vehicle. Proximity sensors, optical filtering, and other space-developed methods are incorporated into a six-passenger experimental tram that automatically follows a thread-like cable on the roadway surface. The low speed tram is started and stopped at will by the passengers themselves. Proximity sensors halt the car when anything comes into its path.

high temperature space material, improved circuitry for accurate digital clocks and radios that use much less electricity, and truly bright searchlights and flashlights for emergency uses. Intense searchlights, which can operate from the cigarette lighter receptacle of your car, were spun off from the xenon arc lights developed to simulate the sun in spacecraft test programs. Such a light is 50 times brighter that any car's high-beam headlights, but it weighs only seven pounds. A smaller unit in the shape of a flashlight produces about nine times as much light as the ordinary two-cell version.

Inertial navigation systems used by Apollo spacecraft have now been extensively applied to commercial aircraft. They have logged more that 15 million hours in commercial flight and have proven to be incredibly reliable and accurate. Today, these navigation systems are flown in some 500 commercial aircraft. They are invulnerable to weather and do not depend on radio, radar, or celestial navigation. Every transoceanic flight uses such a device. Not only has the space program helped people to cross the Atlantic safely, but, if it should happen that the flight goes down

into the ocean, the space program may well again come to the rescue. One of the early successes of transferring space technology was the development of a life raft. The original one, which utilized a radar reflective material, was designed by NASA in 1959 to assure that astronauts could be found if their capsules returned off-course. A later design incorporates a radar-reflective canopy which is colored speckled orange for easier sighting. Within two years, travel safety will be further enhanced. An international effort among the U.S., U.S.S.R., Japan and European nations will soon launch satellites to provide worldwide coverage of emergency transmitters used on downed aircraft and distressed ships. An SOS will be detected instantly and relayed to search and rescue teams.

I think it is fair to say that the spinoffs from space technology have found applications in all modes of transportation, both personal and commercial. The introduction of the Space Transportation System will surely multiply those applications many-fold because of the increased technology and the very nature of the demands upon the system. I look forward to enjoying new innovations in mobility.

YOUR HOME

Your home should be a prime beneficiary of space technology. New building materials, better uses of existing fuels as well as solar energy, fire prevention techniques and tools, and a variety of other household products have all resulted from our national investment in space program research and development. The most significant opportunity for change in houses over the next few decades will be related to energy conservation and management. Our homes consume about 20 percent of the energy used in this country each year. NASA administrators believe that a proper way to demonstrate some of the spinoffs was to construct an actual house, utilizing as many of the new developments as possible. This was done at the NASA-Langley Research Center near Hampton, Virginia. The demonstration house is called The Energy Conservation House and is unlike past "houses of the future." Cost effectiveness is emphasized, and developments expected to be commercially available in the 1980s were in-

cluded. The 1500-square-foot building contains a living room, kitchen-dining area, three bedrooms, two bathrooms, a laundry, a garage, and an outdoor living area. It was occupied by a family for one year and then opened to the public.

The design makes maximum use of energy savings. Heating is provided by solar collectors and a nighttime radiator system using a heat pump. The house even partially reclaims waste water. Many of the energy saving advantages of this house were achieved by the simple fact of good designing. For example, the long axis of the rectangular house is oriented in an east-west manner with large south-facing glass areas. The garage is positioned to protect the house from the north wind. Fireplaces are provided with glass doors to reduce heat loss through chimney flues. Even the landscaping is designed in accordance with the objectives of the house. Another housing development to which space technology has contributed is the unique geodesic dome originally designed by the well-known architect R. Buckminister Fuller.

A new lightweight, inflatable, non-tipable radar-reflection life raft was developed by NASA for recovery of astronauts. The U.S. Coast Guard has adopted these for regular use, and they are also available commercially.

NASA supplied him with a grant to develop these domes into future space structures. Today, large domes are enclosing living areas, swimming pools, tennis courts, and a variety of other commercial work areas. Nine of these structures can be seen at the Kennedy Space Center where they house the U.S. Bicentennial Science and Technology Exposition.

Hopefully we will have more efficient and safer houses in the future. Nevertheless, there is always a chance of a fire in a home. Space technology may again come into play if this should ever happen to your house. While fighting the fire, firemen might use axes to provide ventilation. These are slow, however, and today they might use small amounts of a special explosive which can literally cut holes in structures without blowing them apart. This unique cutting charge was derived from the explosive used to separate parts of the Gemini launch vehicle. Further improvements in firefighting include lightweight air tanks pressurized about twice as much as previous tanks. These new tanks, with the harnesses and regulators, provide greater mobility and longer exposure for firefighters.

In addition to new building materials and fire safety techniques, space research has contributed to the development of many products used in our homes today. For instance, we have been able to extend the life of electric motors used in vacuum cleaners, electric shavers, cameras, and so on, through the use of new space lubricants. These are essentially dry substances which were developed first for bearings and motors that had to work in the vacuum of space onboard a variety of earth satellites. Quartz-crystal clocks and watches that have an accuracy of one minute a year have also come from space technology. They were developed for the Apollo lunar missions where timing had to be incredibly accurate. There is a variety of other products with which we come into contact daily. Once you realize the extent to which space spinoffs have affected us, it is hard to imagine what our daily lives would be like without these advantages.

YOUR ENVIRONMENT

In earlier chapters, we have seen that launch vehicles and the Space Transportation System have adverse effects on our environment as they leave the earth. The Shuttle has an additional sonic boom problem as it returns home. However, these effects seem almost trivial when compared to the advantages that we derive from the space program. The fact that we have access to space enhances our capability to protect our environment. The population and industrial explosions which have occurred over the past 200 years have pushed our environment to the extreme. We live in a world of metal and concrete instead of forests and streams. We were built to go five miles per hour, but here we are, going 500 miles per hour. Our bodies were made to eat when hungry; however, we eat by the clock. We must decide whether to govern our environment by our goals or to allow the environment to govern us. Given our present-day constraints, it would appear that we must govern the environment. Nevertheless, we must continue to protect it. Our natural resources are finite, and we must not waste them. Again, spinoffs from space technology are helping to protect and conserve. We can now locate our natural resources from space, warn of storms and fires from space, and detect pollution and improve the quality of air and water from space.

Among the many purposes of our *Landsats,* the land surveying satellites for locating earth resources, is the monitoring of fresh water supplies. Experts analyzing data from these satellites have found that we currently extract fresh water from only about one 100th of one percent of the total global supply. With this much water around, you might ask why we have periodic droughts in some areas of our country. Satellites are promoting better utilization of our water supply by observing large areas on a repetitive basis. Pictures of snow accumulation and possible locations of sub-surface water supplies in relation to cities, irrigated areas, and industrial developments, make future planning more accurate and economical. It turns out that there is 20 times as much underground water in the United States as in its lakes and rivers. Florida alone has more subterranean fresh water than exists in all of the Great Lakes. Water and volume measurements in Florida are now obtained from orbit and used to regulate the release of water from one area to another around the state. Satellites measure rainfall over remote areas. This helps to determine the size and number of the thousands of temporary small lakes in the southwestern United States. This type of information is critical to proper resource management.

Crop identification is now performed

Hardware developed by NASA for astronauts in extravehicular activities and on the moon's surface has been applied to the creation of lighter weight fireman's airtanks. These new backpacks weigh only 20 pounds for a 30-minute air supply, 13 pounds less than the conventional firefighting tanks.

routinely by our Landsats. NASA's computational facilities process the data in order to make more accurate forecasts of harvest. This type of information is becoming more and more critical as the world's food demands rise. Crop surveying is very closely tied to crop destruction by insects and plant diseases. The screwworm is a grub that destroys cattle, poultry, and wildlife in warm regions. It is among the most damaging of agricultural insects. A satellite sensing mission has been undertaken to irradicate the screwworm in Mexico as has been largely done in this country. At one time these worms infested the U.S. from Florida to California and as far north as Nebraska. Our two countries carried on a cooperative effort in a 50- by 100-mile area in central Mexico. This was the forerunner of a much larger program to irradicate the insect throughout North America. Without satellites, 260 additional weather communications links would have to be constructed to

yield accurate environmental data. Continuous, detailed reports on soil, temperature, moisture, and vegetation coverage are required to determine the insects' breeding habits in order to eliminate them. Similar technology may help extend this technique to other insects, such as the tsetse fly in Africa. There are thousands of square miles of Africa today which are unfit for humans or animals because of this insect.

Another dramatic application of space technology to our environment is the improved ability to detect possible disasters such as forest fires, tornadoes, and floods. Forest rangers tell us that knowing where fires may occur and how they might act is almost as important as knowing how to put them out. Sensors originally developed to detect fires on airplanes and spacecraft have been employed with a network of satellite relay stations and computers to provide a fire-index monitor for forests. These sensors check air temperature, relative humidity, and the flammability of the forest floor litter formerly recorded by forestry personnel on-site. This was initially carried out in an experimental program in California. Measurements are converted and beamed to satellites to avoid interference by the mountains. These data are coordinated with wide area photographs, and the combined information is beamed to a NASA facility in northern California. After being processed, the data are sent to forest ranger stations in the appropriate areas.

Space techniques also contribute to controlling air pollution over our cities and in the upper atmosphere. An extremely sensitive spectrometer was recently developed to determine whether gases released from aerosol spray cans were disrupting the earth's protective ozone layer. This instrument was flown at various levels in and above the atmosphere so that scientists could search for hydrogen chloride which is produced by the breaking down of aerosol gas molecules in the stratosphere. They found that these gas molecules were distributed in a layer starting at nine miles and reaching a maximum concentration of almost one part per billion at about twelve miles. These results don't necessarily implicate the use of aerosol gases, since hydrogen chloride is released naturally from the oceans and volcanos, but the level is relatively high now and must be monitored to detect any future build-up. Aerosol cans release freon which in itself does not harm the ozone layer. However, when it reaches an altitude of about 25 miles, the sun's ultraviolet rays trigger the release of chlorine from the freon. This chlorine breaks down the ozone. Millions of tons of freon are produced annually; it takes about ten years for this gas to drift up to the 25-mile altitude. Thus, continuous measurements are needed at all levels of the atmosphere in order to keep track of the effect of freon on the ozone layer. This type of research is also important in determining the pollution effects of supersonic jet aircraft flying in the stratosphere. These contributions summarize just a few of the space technology applications to monitoring and controlling our environment.

WHAT ABOUT THE WEATHER

The weather, "everybody talks about it, but nobody does anything to change it." How many times have you heard that? Well, I can-

not honestly say that we are about to control the weather. Nevertheless, I can say that we are getting better and better forecasts. Every day, somewhere in the world, people are endangered by some weather extreme. We now have a number of weather satellites, high-speed computers, and rapid global communications to help us in forecasting extreme weather situations and averting disasters.

Construction of the first telegraph line in 1884 made it possible to collect weather reports from many places simultaneously and, thus, to track the movements of weather systems. This made it possible to initiate the first relatively scientific forecasts, almost a century ago. The U.S. Weather Service was first established under the Army Signal Corps in 1870, and it became the U.S. Weather Bureau in 1891.

Weather data are collected by many means. Investigators began sending up instruments on kites and balloons in the mid-18th century. However, there was little progress in this direction until the 20th century when aviation activity made it essential to have information on winds, clouds, icing, and turbulence. By 1938 daily soundings were being made with *radiosondes*, balloons with weather instruments and radio transmitters onboard. These instruments, coupled with reports from ordinary weather stations, ships at sea, and pilots of aircraft were used to provide raw data for meteorologists in order to make their forecasts. The development of electronic computers in recent years has made it possible for most of these data to be handled automatically.

It was obvious, early in the game, that photographs of the earth's atmosphere would be very helpful in visualizing weather systems and forecasting their paths. The first experimental photographs showing cloud formations on a large scale were taken by sounding rockets as early as 1959. Such pictures proved to be extremely useful and motivated the later development of weather satellites which could take pictures almost continuously.

The crew of the hospital ship Hope is very thankful for weather satellites. In 1969, this ship, battered by a typhoon in the Indian Ocean, was in danger of sinking. A ham radio operator in Long Island heard the ship's distress call and contacted what is now the National Environmental Satellite Service. They plotted a course that would get the ship out of danger based on satellite pictures. The

NASA's LANDSAT-C, the third in a series of Earth resources observation satellites, was launched in 1978 from Vandenberg Air Force Base. It is shown here undergoing final testing. Natural resources being measured include water distribution, agricultural fields, and mineral deposits.

GOES-A, a high altitude weather satellite launched for use by the National Oceanic and Atmospheric Administration, undergoes launch preparation at Kennedy Space Center. GOES stands for Geostationary Operational Environmental Satellite. It was launched on October 16, 1975 and gives 24-hour weather coverage for the United States while it sits 22,300 miles above the equator.

captain was told to sail due north out of the storm, and the vessel was saved. Today, this service routinely issues weather warnings based on satellite photographs. Current meteorological satellites are evidently more complicated than the first Nimbus or Tiros satellites that were launched from the Kennedy Space Center starting in 1960. The latest satellite called the Geostationary Operational Environmental Satellite beams back an amazingly detailed photograph of the entire Western Hemisphere every 30 minutes, both during day and night. The eye of a hurricane which is only 50 miles across can be easily seen in these pictures, taken from an altitude of 22,300 miles. There are a variety of other meteorological satellites which assist in gathering world-wide weather data.

VIA SATELLITE

All of us have seen international events "live" on television, but few notice the words "via satellite" periodically flashed on the bottom of the screen. Fewer have though about the meaning of this mini-message. It means that space technology directly influences our lives much more than most of us realize. No single service has affected our lives more than communications. You saw in Chapter 4 how voice, picture, and data links provide full-time contact with the Shuttle while in orbit. This is just an infinitesimal part of our overall communications services available today.

Arthur C. Clarke envisioned a worldwide network of communications satellites in a 1945 article. He described the system which carried telephone and television through orbiting relays at an altitude of 22,300 miles above the earth. This altitude was selected because satellites at this height take exactly one day to make one cycle of their orbit. Earth takes exactly one earth-day to rotate about its axis. Thus, these satellites would move around the earth in synchronism, never rising or setting but maintaining the same longitudinal position. Of course, in 1945 we did not have the technology or the hardware to build and to send up satellites. Fortunately, we did have the goals to reach for and the vision to make the necessary commitments.

The Echo I was to satellite communications what the Wright Flyer was to aviation. What could be simpler than a 100-foot diameter balloon orbiting the earth at low altitude. Its surface was metallized so that signals could be

bounced from one earth station to another. This was done successfully for the first time on August 12, 1960. Shortly afterwards, a series of active, low altitude relays were developed. These included TELSTAR, Courier, and Relay. Each assisted in the further advancement of necessary technology and launch vehicles, satellite design, and ground station development.

Arthur Clarke pointed out in 1945 that there were fundamental advantages to operating a communications satellite at 22,300 miles which were not available at low altitude. For example, ground stations can track low-flying satellites for only a few minutes at a time as they quickly pass overhead. A single, high-altitude satellite can be seen continuously from 42 percent of the earth's surface. Thus, three satellites can provide full-time, worldwide communications coverage, except for the extreme polar regions.

In 1963, the first satellite of this kind was placed into orbit over the Atlantic Ocean. Another one was inserted into orbit over the International Dateline the following year. This was used to relay the 1964 Olympic Games from Tokyo to the United States. We have certainly come a long way in just twenty years. There is no doubt that global communications have been revolutionized by satellites. A three-minute telephone call from New York to London now costs about $3.20. This same call was $12.00 before the launching of the first commercial communications satellite, Early Bird, in 1965. The International Telecommunications Satellite Consortium has a fleet of spacecraft hovering over a number of longitudes around the world, relaying messages over the Atlantic, Pacific, and Indian Oceans. They provide service to 101 countries.

Satellites are no longer limited to relaying special events from overseas. Today they are being used by cable television to reach viewers in remote areas such as the Alaskan outback and offshore drilling rigs. New communications satellites are being launched almost monthly by individual countries and companies for a variety of applications. It may be surprising to realize that individual countries can justify the use of communications satellites stationed more than 22,000 miles above the country, but there are good reasons for this. For example, many countries have broad, territorial expanses and a land-based system of communications lines that

Hurricane Anita is one of nature's fascinating creations. On September 1, 1977 this picture was taken by a GOES weather satellite. The next day, Anita travelled westward into northern Mexico and was dissipated. Note the eye of the hurricane to the left of its central portion.

Here's a view of Earth from the GOES satellite taken on October 25, 1975. North and South America are clearly visible. In fact, there appears to be a storm just east of the Florida peninsula.

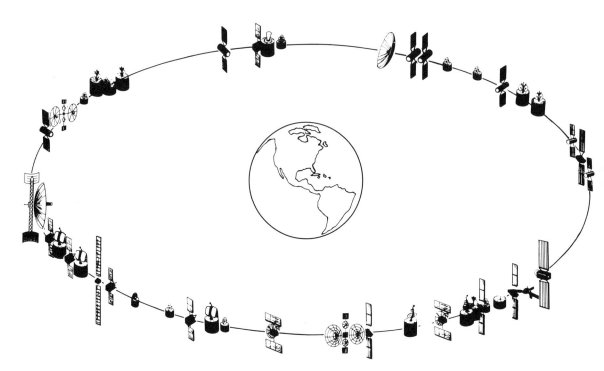

The geostationary orbit is extremely popular. This cartoon illustrates only part of the number and types of satellites seeking positions in this orbit since 1970.

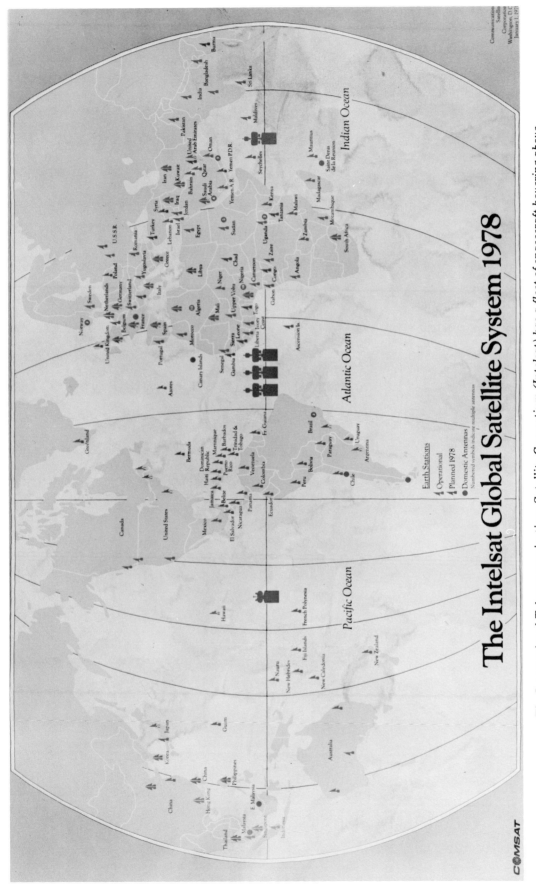

The Intelsat Global Satellite System 1978

Earth Stations
- ⬩ Operational
- ⬩ Planned 1978
- ● Domestic Antennas
  Numbered symbols indicate multiple antennas

The International Telecommunications Satellite Consortium (Intelsat) has a fleet of spacecraft hovering above the equator to service 101 countries.

are extremely expensive to maintain. These satellites can relay as many as 10,000 telephone calls at a time.

In 1961, John F. Kennedy invited "all nations to participate in a communications satellite system in the interest of world peace and closer brotherhood among people throughout the world." That action was the beginning of what we now know as the International Telecommunications Satellite Consortium, the most extensive commercial communications system based on the use of satellites. This consortium, formed in August 1964, is managed by the Communications Satellite Corporation in Washington, D.C. but includes over 90 member nations with more

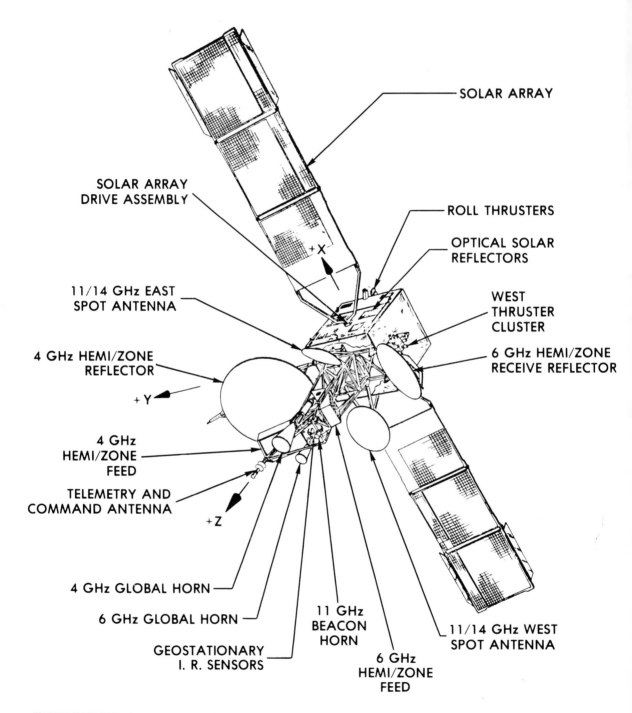

SOLAR ARRAY

SOLAR ARRAY DRIVE ASSEMBLY

ROLL THRUSTERS

OPTICAL SOLAR REFLECTORS

+X

11/14 GHz EAST SPOT ANTENNA

WEST THRUSTER CLUSTER

4 GHz HEMI/ZONE REFLECTOR

6 GHz HEMI/ZONE RECEIVE REFLECTOR

+Y

4 GHz HEMI/ZONE FEED

TELEMETRY AND COMMAND ANTENNA

+Z

4 GHz GLOBAL HORN

6 GHz GLOBAL HORN

11 GHz BEACON HORN

11/14 GHz WEST SPOT ANTENNA

GEOSTATIONARY I. R. SENSORS

6 GHz HEMI/ZONE FEED

INTELSAT V is the latest operational international communications satellite to carry world-wide television, telephone, and other data.

than 110 ground stations. Today, the operational global system is composed of two satellites over the Atlantic, one over the Indian Ocean, one over the Pacific Ocean, and one spare satellite.

Many years ago the United States realized the advantages of augmenting a well-developed, ground-based communications system with satellites. These spaceborne relays would be for domestic uses only. After years of debate about who ought to be allowed to operate such a system, Western Union was allowed to be the first to proceed. Its first spacecraft was launched in April 1974 and called Westar. A second was launched six months later. Since then, the Federal Communications Commission has also authorized American Telephone and Telegraph and the Communications Satellite Corporation to establish a United States domestic system exclusively for telephone traffic carried by the Bell System. The first satellite for this purpose was called Comstar and was launched early in 1976.

I was fortunate enough to be at the satellite control station when RCA Americom launched their first domestic satellite in December 1975. This spacecraft provided video distribution, data, and telephone service for Alaska and the lower 48 states. The number and types of such satellites continue to increase.

There is another customer for communication satellites that we have not yet mentioned. This, of course, is the military. It has its own defense satellite communications system. Its purposes and functions are understandably different from those for commercial applica-

tions. An obvious benefit of satellite use is that the earth terminal can be small and mobile. The military is certain to continue exploitation of satellites for communications, navigation, and data transmission. Defense satellites of this type have already proven themselves in several incidents around the world.

WHAT NEXT

So far, we have only touched upon a few of the benefits derived from the space program. Some of these have been indirect, such as those in the health field. Many have been direct benefits, for example, communications satellites. It would be extremely easy to fill a whole book with the spinoffs and applications of space technology.

I hope by now that you will agree that our lives have been changed significantly due to the relatively small investment in space research and technology. Will we continue to derive benefits from this work? I think the answer is definitely yes! We can expect much more efficient transportation, better conservation of energy, more efficient utilization of our resources, and healthier and longer lives. All this will be the result of an active and well-funded space program

How does the Shuttle fit into all this? It will provide the vehicle for transporting most of our space payloads into orbit between 1982 and 1995. Thus, it will be the workhorse of the space program. We are about to look beyond the Shuttle to try to glimpse the wonders that we might expect as the year 2000 approaches.

Large space structures would be too bulky to carry up in their operational form. Thus, structural elements, such as beams, will be fabricated in orbit. Some will be extruded directly from equipment in the Orbiter cargo bay.

Future space stations may also be construction bases for solar power stations. Here, we see one of these power stations being built by a crew that lives in a cylindrical base.

CHAPTER SEVEN

TOWARD THE YEAR 2000 AND BEYOND

PRELUDE TO THE FUTURE

The stage is set. The Space Transportation System has become a reality. Our past history has shown the potential of space. We are about to experience another technological revolution. This time the "high frontier" will be conquered. Space will, someday soon, no longer be reachable only by a handful of astronauts. The Space Transportation System will provide the bridge to span the gap between scientific and societal missions; between highly expensive and commercially economical flights; and between sending up superbly conditioned crews and average people.

The Space Transportation System will do more than simply deliver payloads to orbit. It will be a platform from which new industries can be generated; from which new advances in life sciences can be developed; and from which new sources of energy can be harvested for use on earth. Let's try to visualize some of the events of the 1980s in order to anticipate the coming space revolution.

There is a consensus in the space community that large and complex structures will be orbiting our earth sometime in the future. In-space construction, assembly, and maintenance will, of course, be necessary. The attraction of space is its view of earth, weightlessness, and unlimited pollution-free energy and heat dissipation. In 1976, NASA

commissioned initial studies of a "space construction base," not to be confused with the "space station" concept. This new approach emphasizes the manufacture of a space base in orbit, rather than simply a laboratory in which experiments are carried out. An early version could be operating as soon as 1984. Starting with a small crew of four to eight people, it could grow considerably. A number of Shuttle flights would carry modules up to be joined in orbit. It turns out that potential users of this space base want large-scale space structures, many times bigger than the Shuttle; however, the scaling-up process is going to be an evolutionary one. Remember, the Skylab sunshade is our biggest space construction project so far.

The Orbiter will provide a platform from which we may assemble early space structures, but this vehicle is limited by expendables related to electric power and attitude control. *Fuel cells* are its prime source of electricity on orbit. These use super-cold liquid oxygen and liquid hydrogen which are stored in cryogenic tanks. They can provide a total of seven kilowatts of power for up to 30 days but require about a quarter of the payload capacity for the cryogenic tanks. A "power module" is being developed to augment the Orbiter's stay time aloft. This unit will be equipped with large solar cell arrays

The deployment of the parasol sunshade on Skylab in 1973 represents the only space construction experience the United States now has.

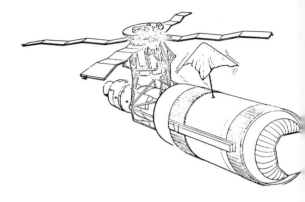

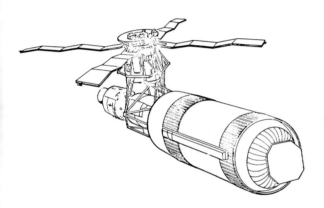

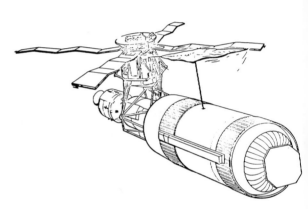

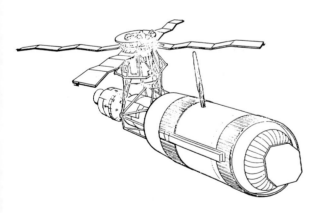

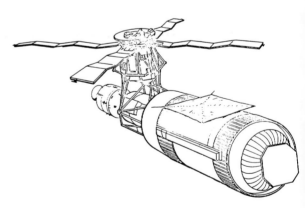

and gyroscopic attitude control system. Its functions are to augment the Shuttle's available power and extend the stay time in orbit. The module will remain in the payload bay, but it will deploy radiators and solar arrays while in operation. A modest size unit that provides 25 kilowatts to the Orbiter can increase the stay time to 60 days. Such a device should be ready in the mid-1980s.

The power module concept may well find a variety of uses. It can provide power to free-flyer payloads by making minor modifications to its design. Such devices may be thought of as orbiting power stations at which payloads can be "plugged in." Spacelab could be a primary user of these modules. This combination of staffed and unstaffed devices could be a long-duration, staffed free-flyer.

By the late-1980s there will be many missions which require much higher power levels than will be available from the power module. Over 100 kilowatts will be needed for applica-tions such as advanced communications satellites which provide electronic mail to individuals; for high powered radar systems which track movements of vehicles and vessels and other defense applications; and for testing of communications between space and under-the-ocean sites. How can we provide this kind of power in orbit? There are basically two ways: nuclear reactors or large solar collectors. NASA has been studying and comparing these two possibilities for some time. So far, the solar power approach seems to be favored. Solar energy, when allowed to fall onto a collection device, can either be directly converted to electricity through the use of light-sensitive cells or can be used in other devices for later conversion to electrical power.

Several configurations of a 250-kilowatt power module have been conceived. All of these are large and require a space construction capability. A number of new innovations are needed to get ready for these large space

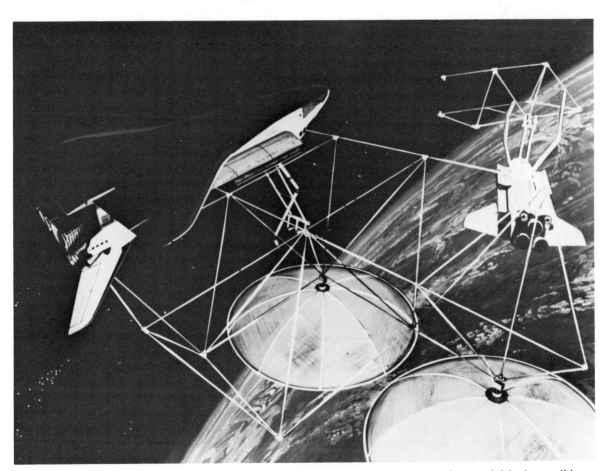

The Space Shuttle will enable us to take advantage of the space environment to perform activities impossible on the Earth's surface. Assembly of giant antennas that will provide low cost, personal communications services for millions of people around the world is just one example. Benefits could include direct home broadcasts of television programs, electronic mail, search and rescue operations, health care, and person-to-person phone calls anywhere in the world.

Large space structures appear to be in our future. Here is an artist's concept depicting activity at a possible staffed, modularized space station in Earth orbit. The modules, which house various equipment, functions and activities of the space station, could be carried to orbit by the Space Shuttle.

A beam builder is at work fabricating a large structure in Earth orbit as a development step for a large solar power station. The Space Shuttle would serve as a means of transportation and a work base for the construction.

projects. We will have to develop free-flying teleoperators or robots which can carry out assembly operations. New tools, jigs, and fixtures are needed to make assembly and construction as fast as possible. To manufacture structures, we will need fabrication equipment to form metal parts from sheet stock. This is something like forming aluminum rain gutters at the site of residential housing construction. Space fabrication has several advantages. Materials may be carried to orbit in bulk form. Eventually, extrusions of very long structural elements will be possible. Such components will be needed for a variety of proposed concepts. The very task of constructing a 250-kilowatt power module will allow us to develop a new capability for in-space building.

Let's take a brief look at our reasons for building large structures in space. We, on Earth, accept gravity and the atmosphere without question. Until recently, you would not have thought there were advantages to escaping these environmental realities. However, a whole new industry is evolving due

to our capacity to travel to and from orbit: *space processing*. The cosmos, with zero gravity and almost perfect vacuum, offers prospects for large-scale manufacturing of products which are difficult or impossible to make on Earth. Thus, *space processing* is the exploitation of those unique features of space for manufacturing purposes. Benefits of this new industry include lower costs because of more efficient processing and the capability to produce materials or devices not otherwise available.

Atmospheric effects can be controlled to some extent on the ground, but gravity cannot be eliminated. Gravitational effects are severe, especially when processing a mixture of material of different densities and temperatures. On the ground, these materials tend to segregate, denser substances float to the bottom and lighter ones end up on top. Even if a liquid is a pure substance, internal currents are caused by temperature differences, because higher temperatures mean lower densities. In space, mixtures stay mixed. This permits the solidifying of substances in suspension in uniform distribution without separation. An additional advantage is that liquids can float freely without containers. This avoids any reaction or deformation caused by walls or edges.

Without orbital flight, the effects of gravity can be controlled for only a few seconds or minutes. Drop towers permit a few seconds of freefall. Aircraft nose-overs allow up to 30 seconds of reduced gravity. Sounding rockets can produce a few minutes of weightlessness. Apollo and Skylab were the first missions to offer experimental opportunities for extended periods. The Space Transportation System will continue these opportunities. So far, only a few experiments in *space processing* have been done. Nevertheless, these indicate that the potential return is great, and the framework for a future industry has been established. The 1980s will see hundreds of additional experiments as we witness the dawn of a cosmic industrial revolution.

The future will certainly bring more and more staffed adventures into space. Thus, many Shuttle activities of the 1980s will be concerned with life sciences. More knowledge about long exposure of the human body to weightlessness is needed for staffed space probes and orbiting colonies. This same research will tell us more about life processes in general. The conditions of space can also be used to isolate, purify, and synthesize biological materials such as antigens, hormones, and antibiotics. This is a part of a continuing effort to improve health care here on earth.

Long before the first Mercury flight, serious questions were raised about the physiological effects of space. There was so much concern, that our first astronauts were little more than passengers on automated spacecraft. From the first 15-minute, 22-second Mercury ballistic flight to the 84-day Skylab endurance test, exposure to weightlessness has been increased in deliberate steps. We are now confident that human beings can adapt and function effectively for extended periods while in orbit.

In the Mercury program, medical measurements indicated that normal body functions were not adversely affected. The Gemini program allowed medical studies of crewmembers in space for up to 14 days. This was also our first opportunity to demonstrate the usefulness of the on-board crew in the weightless environment. Skylab allowed the first overall picture of how a person lives in space. The preceding chapter pointed out that medical monitoring methods, life science experiments, and engineering techniques used in these orbital missions have added greatly to our earth-bound medical and biological research and clinical practices. In the coming years, we will see advances beyond our imaginations. These will include fields such as cellular and molecular biology, radiobiology, botany, zoology, and cardiovascular functions.

Earth is a spaceship with limited energy. We don't normally think of our mother planet in this way, but our energy crisis is forcing us to take a new look at what we have. Yes, earth is a spaceship, a vehicle whose internal energy is running low. We must look to new power sources. There are really only two choices for the coming decades. One is nuclear power here on earth. The other one is power beamed down from space. Where does this heavenly power come from? There are again two possibilities. Nuclear reactors could be sent into orbit to generate this power, which is then transmitted to earth. This would relieve some of our concerns about potential hazards and pollution on the ground. However, the more likely source in orbit is the one earth has used since its creation, the sun. You may be surprised at first that anyone would propose using solar power from orbit. The earth does a good job of collecting it but not as good as orbiting solar power stations. First, this is a non-polluting, non-depletable energy source. Solar power plants on earth suffer from dilution of energy as it travels through the atmosphere, and the sun only illuminates about one-half of the globe at a time.

Optimum positioning of such power plants in orbit can insure unfiltered sunlight almost without interruption. These systems should be as much as 20 times better collectors than ground units. Satellite power stations would probably be placed into the same orbit as our communications satellites, at an altitude of 22,300 miles, again taking advantage of the constant longitude property. Enough electricity could be produced to power a large city by a single satellite containing several square miles of solar collectors. This power would be converted into microwaves and be beamed precisely to earth ground stations from which it would be distributed to consumers. These microwaves are very similar to the form of energy used in our modern radar ovens. Pollution and hazard effects appear to be minimal. The primary concern is long-term exposure of human beings, animals, and vegetation to microwaves. However, any detrimental effects would be confined to a very small area near the receiving ground stations.

Technology to design these power stations appears to be within our grasp. Two major problems remain to be overcome: the cost associated with transporting materials and tools to high altitude orbits and the implementation of assembly in orbit. Economic feasibility has not been demonstrated and probably will not be for several years to come. It appears that orbiting power stations will be economical only if we have a low cost Space Transportation System, much lower than that of the current Shuttle design. Thus, a second generation transportation system will be necessary if we are to launch components of the solar power stations from earth. One alternative recently proposed is that we use materials from the moon's surface to build these orbiting power satellites and have manufacturing facilities which are also in orbit. We have yet to develop the assembly strategy and capability for building such large structures in space, however.

A full-size solar power station would remain fixed over one position on the Earth and continuously beam power via microwaves to the surface. This huge space structure is envisioned as being several miles in length and width.

169

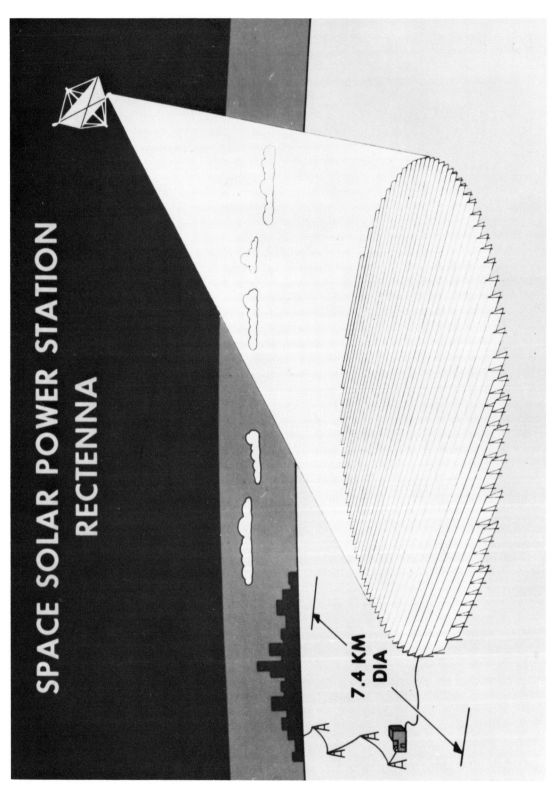

Here is an artist's concept of the ground-based microwave receiving station which would be used in connection with the solar power station orbiting high above the Earth. This ground station could be located near the consumer, and be largely independent of weather considerations.

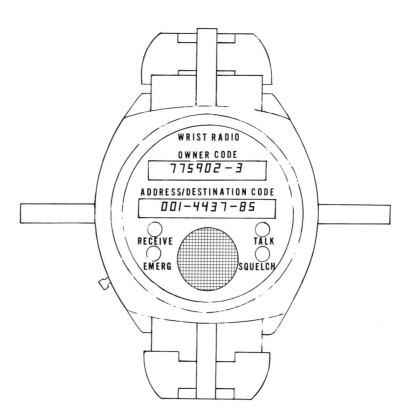

We may someday soon be using Dick Tracy wristwatch radios like we now use telephones. Huge satellites will relay your voice to any spot on Earth.

DICK TRACY WRISTWATCH RADIOS AND MORE

Let's take a peek into what space might bring us over the next 20 years. This time, though, we will concentrate on those innovations which could change our daily lives and the way we do things. Keep one thing in mind; what you are about to read contains a fair amount of speculation without consideration for economic and societal implications. These ideas represent reasonable extensions of what we know today and are not just science fiction concepts.

Let's begin with the topic of communications. Today, a single communications satellite provides coverage over nearly an entire hemisphere. However, operations are limited to relaying information between specific ground terminals, each employing large-size antennas and high-powered transmitters. Thus, the satellite itself can be relatively small. In recent years, satellite designers have tended to increase their size and power so that links could be established between aircraft in flight, small ships, and portable ground terminals. Looking to the future, this trend may well continue to the point where *personal communications* are possible. It is not too difficult, then, to imagine the construction of large orbiting communica-

tions satellites, capable of providing communication links between a great many wristwatch size terminals on the ground. Thus, we may well see the Dick Tracy wristwatch radio of yesterday's comic strip become a technological reality. Just think of it; everyone would have a telephone strapped right to their wrist. It could be similar in size, weight, and appearance to today's digital wristwatches. The entire unit could weigh less than two ounces, and mass production methods could probably bring the price down to within everyone's pocketbook.

We can imagine many other uses for the personal communications system. Wrist radios are ideal for police work. A patrol officer need not be tied to a vehicle to keep in touch with the home station. Police signals could be automatically encoded so that jamming would be very difficult. Therefore, even the smartest criminals could not interfere with the use of these small radios. Furthermore, police messages could be easily scrambled to avoid interception by the wrong parties. Wrist radios could also be used for instantaneous polling by the government on major issues. They could even be used as portable voting booths in elections. Needless to say, the voter turnout would be extremely good. Just imagine over 100 million people voting within a single hour.

We are all familiar with the recent history of the U.S. Postal Service and its difficulties in making a profit. Think about this possibility for a moment. Instead of sending mail in its familiar form, the information content could be sent by communications satellite. Letters collected in the conventional way by the originating post office could be read by automatic television scanners and transmitted electronically to a receiving post office. Here, a facsimile would be printed and delivered to the addressee by ordinary mail carrier. Actually, we already have something like this, called the *Mailgram*. So far, this is a combination telegram and letter, but, eventually, it may evolve into *electronic mail*.

A very similar satellite could be used to disseminate data to small users within the United States. From one's home or office, a terminal could provide access to libraries and other information sources. Only a relatively small antenna of about three feet in diameter would be required which could be permanently mounted within your attic space or on the roof of your office building. This concept could include direct broadcast television to your home or office. Broadcasting stations from anywhere in the world could send programs via satellite directly to your receiver. We will have the first direct broadcast television by 1986.

Many of us who travel frequently to business meetings find it expensive, time consuming, and disruptive to home life. By applying *holography*, the science of projecting three-dimensional images, it could be possible to send images to a meeting instead of sending people. Imagine a conference room fitted with a multi-color illuminator. A television camera picks up a holographic image and relays it via satellite to a set of projectors in a second identical conference room across the country. Three-dimensional images of people and objects at the transmitting location are projected in the receiving room. These images appear completely lifelike in color and in perception. They can move, speak, and otherwise be a true replica to the observer. In the recent movie, "Future World," holograms were used to animate a chess game. This is almost as good as the *transporter* used on "Star Trek."

Still another satellite application might be the protection of our borders. Here, a group of intrusion sensors are placed around a property or along the border. The sensors might detect the sound of footprints, the breaking of a special electronic beam, the opening of an electronic switch on a door or window, or the pressure of an intruder on the floor. Once the sensor is activated, it radiates a coded signal to a high-altitude satellite. This is then relayed to the appropriate security control center where police or border guards can take action. The real advantage of using a satellite in this application is its wide-area coverage. Sensors from the entire continental United States can be monitored simultaneously by a single satellite. Thus, our four borders can be secured through the use of such a space platform.

The few examples offered here represent only the "tip" of the iceberg. There are an uncountable number of other potential applications for large earth orbiting satellites. These are related to things like personal navigation, international air-traffic control, monitoring of energy distribution around the world, and night illumination of our cities. Today, many of these things seem far fetched. However, remember that just 25 years ago nobody believed that we would be on the moon by 1970.

SECOND GENERATION SHUTTLES

The Space Transportation System marks the end of an era in space. It represents the end of the expendables in space flight. More important, it is the beginning of a new age in the utilization of the cosmos. It is a first step to a very prosperous future for all of us. The Space Shuttle is designed and intended for use until the mid-1990s at which time it should be replaced by its successor. What will its replacement look like, and how much more will it do? We have now seen some of the possible applications for such a system in the areas of large space structures, life sciences, and space power stations. Let's speculate on this new system.

During the next decade, technology will most certainly continue to surprise us. It should allow the development of more economical transportation to space by the mid-1990s. For example, light structures and more efficient rocket engines contribute to this objective. We are already starting to gather the required technology because experience tells us it takes as much as ten years to initiate and to carry out programs which advance our knowledge sufficiently for a new system of this magnitude. Another six or more years may be needed to design and build the

PERSONAL COMMUNICATIONS

- 150 ft DIAMETER ANTENNA
- 25,000 SWITCHED CHANNELS

- WRIST RADIO COST = $10
- MORE THAN 2,500,000 USERS

Here are a few example uses of personal communications via satellite. With 2.5 million users the cost of each wristwatch radio might be as low as $10.

BORDER AND FACILITY SURVEILLANCE

- 1 MILE LONG ANTENNA
- ONE GEOSTATIONARY RELAY COMSAT

- 1,000,000 SENSORS
- 1 POUND EACH, 10 YEAR LIFE

Another satellite application might be the protection of national and property borders. Sensors are placed along the perimeter or boundary to detect sounds or the interruption of an electronic beam. Activation of a sensor results in the transmission of a coded signal to a satellite and on to the appropriate security control center.

The Space Transportation System is opening a new era. On the left are depicted definite projects now underway. As we move to the right the more exotic and long-range space projects are shown.

components and to get it into operation. Thus, a new Space Transportation System, ready for 1995, will require work to begin now. NASA has, in fact, begun work on the next generation Shuttle. Initial studies have assumed the same payload capability with a *single-stage-to-orbit* vehicle. In other words, the system would be fully reusable and use only one craft from liftoff to landing. There would be no solid rocket boosters or external tank. The main rocket engines might use liquid oxygen and liquid hydrogen as does the present Shuttle. Advanced engines are also being considered.

There are actually several possible approaches to designing a second generation system. One would concentrate on simply upgrading the present design. This has been successful with expendable launch vehicles in increasing payloads by as much as 30 percent. Potential upgrading includes weight reduction, better booster propulsion, and drag reduction during ascent. Another approach is to develop a fully reusable system with a flyable first stage that returns to the launch site. The present Orbiter and external tank would be replaced by a fully reusable Orbiter with onboard propellants. This approach would require the development of two new vehicles. A third approach is to develop a whole new transportation system. The outcome of this would probably be the single-stage-to-orbit configuration, starting from a new set of mission requirements and designing from the latest technology available.

The correct path to our next generation Space Transportation System is not at all clear. We can say, however, that the choice will depend on available resources, demand for Shuttle services, and new technology. Some of NASA's future predictions on space activities indicate that a single Space Transportation System may not be sufficient for all missions. We could someday have two dif-

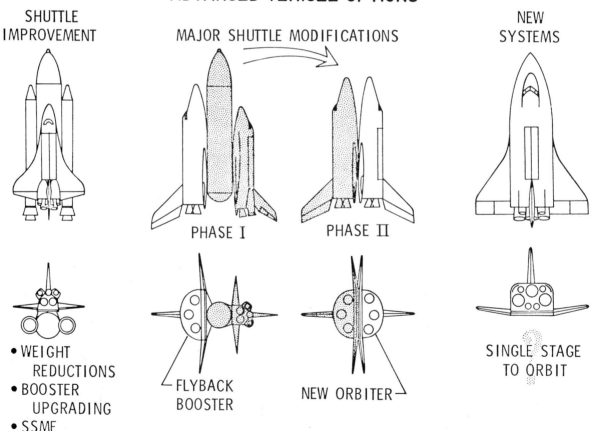

ADVANCED VEHICLE OPTIONS

SHUTTLE IMPROVEMENT

MAJOR SHUTTLE MODIFICATIONS

NEW SYSTEMS

PHASE I   PHASE II

- WEIGHT REDUCTIONS
- BOOSTER UPGRADING
- SSME UPGRADING

FLYBACK BOOSTER

NEW ORBITER

SINGLE STAGE TO ORBIT

Second generation Shuttles will be developed through one of three methods. The present design could be upgraded. A fully reusable system with a flyable first stage that returns to the launch site is another possibility. Finally, a whole new Shuttle may be developed which uses the concept of single-stage-to-orbit.

An artist's conception of O'Neill's original space settlement idea shows two huge cylinders with long thin mirrors to control the cycles of day and night. Industrial activities are located on the axis of the cylinders outside the pressurized area. This permits low or zero gravity operations. Rings of independent agricultural growing areas allow seasonal phasing to optimize harvesting.

ferent systems. It will certainly be interesting to see how the next generation Shuttle takes shape.

HUMAN COLONIES IN THE COSMOS

What do we see beyond the year 2000? The only large projects now being contemplated for the far future are the space power station, manufacturing facilities in orbit, and space colonies. We have already touched upon power stations which beam energy to earth on microwaves, and space manufacturing. Let's look at human colonies in the cosmos. Speculation about staffed spaceflight goes back almost 2000 years. In the last century the concept has become more interesting. For example, in 1895 the famous Russian scientist, Konstantin Tsiolkowsky, described a space station which would use artificial gravity and solar energy. In 1952, Wernher von Braun was proposing space stations in this country. Arthur Clarke published a novel about that same time which involved large orbiting stations. The list goes on and on. However, early proposals were largely connected with science fiction and little engineering work was done.

Several well-known people have suggested the use of extraterrestrial resources; for example, minerals from the moon and asteroids. The rocket inventor, Robert Goddard, suggested this in 1920. In 1950, Arthur Clarke noted the possibility of mining the moon and launching this lunar raw material by electromagnetic accelerators along a straight track on the surface. This technique is analogous to the principle which makes electric motors work. In this case, the electric motor runs in a straight line rather than rotating. Carl Sagan, the famous researcher on extraterrestrial life, suggested in 1961 that we inject colonies of algae into the Venusian atmosphere to reduce the concentration of carbon dioxide around that planet. About that same time, Freeman Dyson proposed the processing of materials from uninhabited planets and moons to construct habitats in orbits about the sun. The most recent revival of these concepts began in 1969 with a series of detailed studies conducted by Gerard O'Neill, a physics professor at Princeton University. This work has captured the interest and imagination of millions. A midnight lecture at a recent *Star Trek* convention drew over 10,000 people. Even a significant part of the scientific and engineering community has become involved. NASA has modestly supported his work in order to allow objective evaluation of its feasibility.

History tells us that an idea which is ahead of its time, periodically reappears with ever greater interest and enthusiasm until it can be implemented or is superceded by a better idea. Has space colonization's time come? We cannot yet answer this, but let's take a look at the most recent version of this old idea. The first of many articles and literature on O'Neill's work appeared in the September 1974 issue of *Physics Today*, with the ambitious title, "The Colonization of Space." He admits, "... initially almost as a joke, I began some calculations on the problem in 1969, at first as an exercise ... As sometimes happens in the hard sciences, what began as a joke had to be taken more seriously when the numbers ... come out right."

He claims that his early studies lead to several important results: we can colonize space, and, without robbing or harming anyone or anything. All industrial activities could be moved away from earth within a century. The migration of people and industry into space is likely to encourage independence, small-scale governments, and cultural diversity. The ultimate size of the human race in space could be 20,000 times its present value.

How does one go about colonizing space? O'Neill suggests that it is possible with existing technology. Several elements are needed in order to live normally: energy, air, water, land, and gravity. As we know, solar energy is abundantly available in space and convenient to use. The moon and asteroids could supply the needed materials. Artificial gravity can be induced by rotating the colony. The true habitat would be self-sustaining, thus, producing its own food and oxygen. The various requirements of man seem to dictate a geometry which will give normal gravity, a day and night cycle, natural sunlight, and earth-like appearance. These conditions lead to a configuration consisting of a pair of cylinders. A minimum size is dictated by the requirement that it should be self-sustaining. It turns out that each cylinder would be about four miles in diameter and perhaps fifteen miles in length. Within the cylinder land areas are devoted to living space, parks, forests, lakes, rivers, and other things familiar to earth. The circumference of these cylinders is divided into alternating strips of land area, which appear to be valleys, and window areas,

Another view of O'Neill's early space colony concept shows that the space habitat is fully shielded against cosmic rays by a spherical shell made of lunar material. This community could conceivably house up to 10,000 inhabitants. Scientists envision a park-like setting, complete with trees, flowers and streams. Artificial gravity is produced by rotation of the cylinder about its axis.

which allow sunlight to enter. Each of these two habitats rotates about its axis every two minutes, with the axis always pointed toward the sun. Agricultural activities are carried on outside of the cylinders, in a series of smaller units.

The agricultural area would be designed to offer the best climate for each crop which is grown. Gravity, atmosphere, and heating are earth-like. No insecticides or pesticides will be needed because the environment would be sterile. All food would be fresh. Growing can be controlled in such a way that seasonal crops can be harvested at the optimum time.

What would it be like in such a colony? There is an abundance of food and clean electrical energy. The climate is controlled to give mild temperatures and unpolluted air. It certainly seems like this might be a very pleasant place to live. The colony is small enough so that bicycles and low-speed electric cars are quite adequate for transportation. Transport between colonies could be accomplished by relatively simple spacecraft. All Earth sports

would be possible within these cylinders. Some of them would be even more fun in space. Imagine mountain climbing with the gravity decreasing as you reach higher and higher altitudes. Since gravity is artificially induced by rotating, the magnitude of this force decreases as you get closer to the axis of rotation.

How realistic are these ideas? In an effort to answer this question, there have already been two conferences on space colonization, one in 1975 and one in 1977. Both were sponsored by NASA and other agencies and societies. These conferences were well attended; participants came from the U.S., Canada, and England. Speakers discussed technical, economic, and cultural problems in establishing space colonies. No insurmountable stumbling block was found to invalidate the concept of a space colony. However, the decision to go ahead with such an adventure would be based on realistic evaluation of costs, schedules, and the potential benefit to humanity.

I think some form of space colonization is

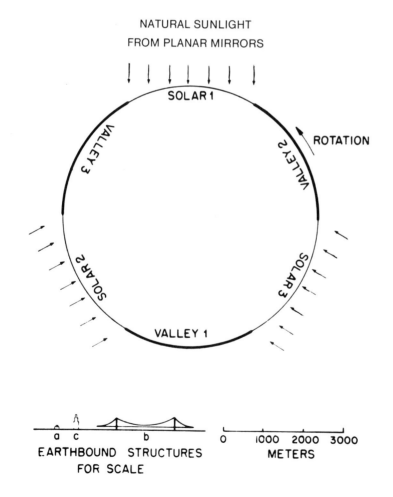

A section of O'Neill's space community shows that the circumference is divided into alternating strips of land areas and window areas. These valleys offer new landscaping opportunities and permit the duplication of certain desirable Earth features.

inevitable. The question is not whether there will be space colonies but, rather, when will they become a reality. We most certainly will not see a true space colony in our lifetime. The present day costs and schedule constraints indicate a program that will take several generations to complete and probably the equivalent of at least one National Budget. I think it is fair to say that the total costs of such a project would be of the order of a trillion dollars; that is, a thousand billion dollars. Can you imagine any politician voting for such a program that would stretch beyond his or her term in office? Furthermore, there are no obvious, real benefits to having space colonies.

They won't solve our population problem here on earth. They won't dissolve our pollution, nor will they completely solve our energy problem. Space colonies might afford a manufacturing facility at which we could build space power stations to assist the earth energy problem, but this will only be a partial help. By their very nature, space colonies will become independent of earth. That means they would be self-sustaining with respect to life supporting elements, with respect to societal needs, and with respect to their own governance. Experience with human nature tells us that we would someday have real *Star Wars*.

Space colonies may well be in our future. A number of concepts have recently been proposed in addition to O'Neill's cylindrical configuration. The one shown here is a toroidal shape, constructed of materials extracted from the moon.

Appendices

APPENDIX A

A Partial List of Companies Receiving Contracts of Less Than $10 Million for Space Shuttle Work

Aerospace Avionics, Bohemia, New York

Aiken Industries, Jackson, Michigan

Cutler Hammer, Milwaukee, Wisconsin

Aircraft Instruments Co., Montgomery, Pennsylvania

Applied Resources, Fairfield, New Jersey

Avco, Wilmington, Massachusetts

Aydin, Vector Division, Newton, Pennsylvania

Ball Brothers Research Corp., Boulder, Colorado

Bendix Corp., Teterboro, New Jersey

B. F. Goodrich Co., Troy, Ohio

Brunswick, Lincoln, Nebraska

Calspan, Buffalo, New York

C. S. Draper Labs., Cambridge, Massachusetts

Chem Tric, Rosemont, Illinois

Collins Radio, Cedar Rapids, Iowa

Corning Glass, Corning, New York

Dynamics Corp., Scranton, Pennsylvania

Edison Electronics Div., McGraw Edison, Manchester, New Hampshire

General Electric, Valley Forge, Pennsylvania

George A. Fuller Co., Div. of Northrop Corp., Chicago, Illinois

Globe Albany, Auburn, Maine

Grimes Manufacturing, Urbana, Ohio

Harris Corp., Electronics Systems Div., Melbourne, Florida

Haveg Industries, Inc., Winooski, Vermont

Honeywell Inc., Minneapolis, Minnesota

Intermetrics, Cambridge, Massachusetts

Jet Electronics, Grand Rapids, Michigan

Life Systems Inc., Cleveland, Ohio

Magnavox, Ft. Wayne, Indiana

Metalcraft Inc., Baltimore, Maryland

Modular Computer Systems, Ft. Lauderdale, Florida

OEA, Des Plaines, Illinois

RDF Corp., Hudson, New Hampshire

Sterling Transformer Corp., Brooklyn, New York

Teledyne Thermatics, Elm City, North Carolina

Tulsa Division, Rockwell International, Tulsa, Oklahoma

Westinghouse Electric Corp., Lima, Ohio

Xebec Corp., Kansas City, Missouri

A Partial List of Companies Receiving Contracts in Excess of $10 Million for Space Shuttle Work

Aeroject General, Sacramento, California

Beech Aircraft Corp., Boulder, Colorado

Boeing Aerospace Co., Seattle, Washington

Fairchild Republic, Farmingdale, New York

General Dynamics Corp., San Diego, California

Grumman Corp., Bethpage, New York

Hamilton Standards Div., United Technology Corp., Windsor Locks, Connecticut

Honeywell Inc., St. Petersburg, Florida

IBM Corp., Owego, New York

Lockheed Missiles and Space Co., Inc., Sunnyvale, California

Marquardt Co., Van Nuys, California

Martin Marietta, New Orleans, Louisiana

McDonnell Douglas Astronautics Co., St. Louis, Missouri

Pratt & Whitney Div., United Tech. Corp., East Hartford, Connecticut

Rockwell International, Los Angeles, California

Sperry Rand Corp., Phoenix, Arizona

Sundstrand Corp., Rockford, Illinois

Thiokol Chemical Corp., Brigham City, Utah

TRW Systems, Redondo Beach, California

United Space Boosters, Inc., Sunnyvale, California

Vought Corp., Dallas, Texas

ECONOMIC IMPACT OF SPACE SHUTTLE

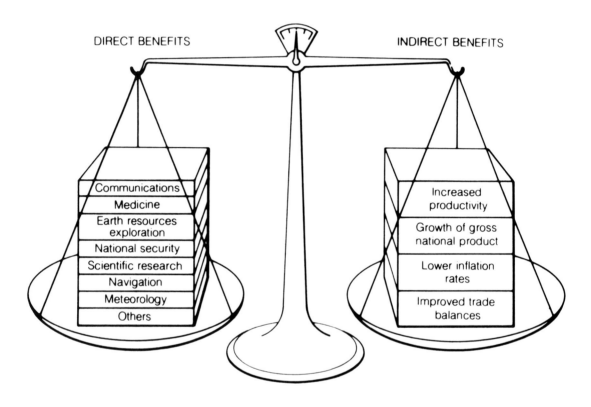

DIRECT BENEFITS

INDIRECT BENEFITS

Communications
Medicine
Earth resources exploration
National security
Scientific research
Navigation
Meteorology
Others

Increased productivity
Growth of gross national product
Lower inflation rates
Improved trade balances

APPENDIX B

SPACE SHUTTLE MODEL INFORMATION AND SPECIFICATIONS

Profile of Shuttle Mission

Each Shuttle orbiter can fly a minimum of 100 missions and carry as much as 29,484 kg (65,000 pounds) of cargo and up to 7 crew members and passenger/specialists into orbit. It can return 14,515 kg (32,000 pounds) of cargo to earth.

SEPARATION OF EXTERNAL TANK

ORBIT INSERTION AND CIRCULARIZATION
HEIGHT: 185 km (115 miles—typical)

ORBITAL OPERATIONS
DURATION: 7-30 days

SEPARATION OF SOLID-ROCKET BOOSTERS
HEIGHT: 43 km (27 miles)

SHUTTLE CHARACTERISTICS
(values are approximate)

LENGTH
SYSTEM: 56.14 m (184.2 feet)
ORBITER: 37.24 m (122.2 feet)

HEIGHT
SYSTEM: 23.34 m (76.6 feet)
ORBITER: 17.27 m (56.67 feet)

WINGSPAN
ORBITER: 23.79 m (78.06 feet)

WEIGHT
GROSS LIFT-OFF:
1,995,840 kg (4.4 million pounds)
ORBITER LANDING:
84,778 kg (187 thousand pounds)

THRUST
SOLID-ROCKET BOOSTERS (2):
12,899,200 newtons (2.9 million pounds)
of thrust each at sea level
ORBITER MAIN ENGINES (3):
1,668,000 newtons (375 thousand pounds)
of thrust each at sea level

CARGO BAY
DIMENSIONS:
18.28 m (60 feet) long, 4.57 m (15 feet)
in diameter
ACCOMMODATIONS:
Unmanned spacecraft to fully equipped
scientific labs

ATMOSPHERIC ENTRY
VELOCITY: 28,082 km/hr
(17,450 mph)

LANDING
CROSSRANGE: ± 2037 km (± 1100 nautical miles)
(from entry path)
VELOCITY: 341-364 km/hr (212-226 mph)

SHUTTLE LAUNCH

SPACE SHUTTLE ASSOCIATE CONTRACTORS

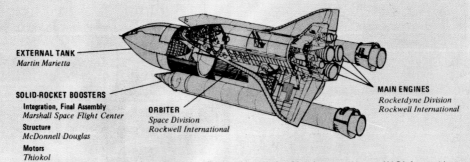

EXTERNAL TANK
Martin Marietta

SOLID-ROCKET BOOSTERS
Integration, Final Assembly
Marshall Space Flight Center
Structure
McDonnell Douglas
Motors
Thiokol

ORBITER
Space Division
Rockwell International

MAIN ENGINES
Rocketdyne Division
Rockwell International

The Space Division of Rockwell International is prime contractor to NASA for total integration of Space Shuttle systems, including all systems being produced by associate contractors.

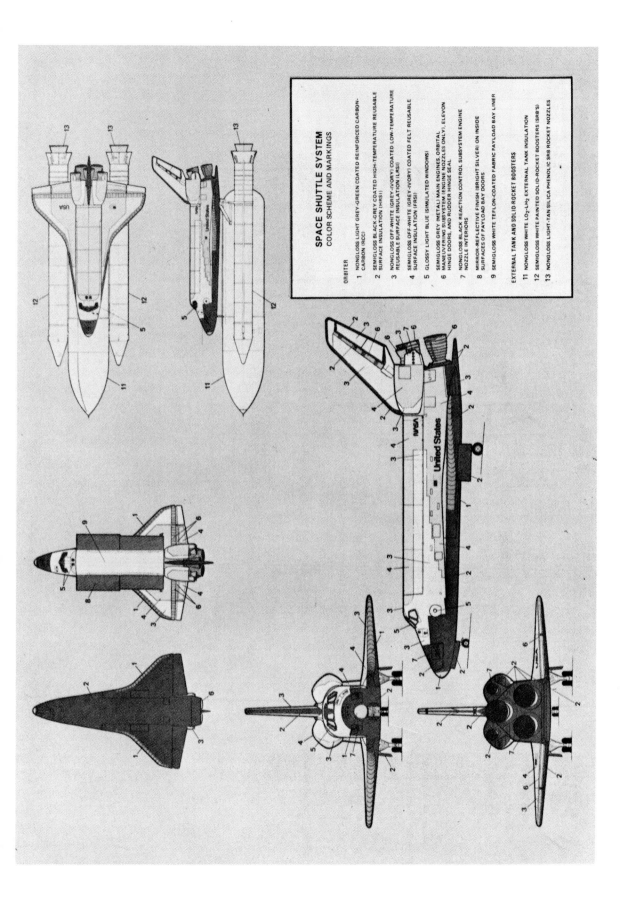

SPACE SHUTTLE SYSTEM
COLOR SCHEME AND MARKINGS

ORBITER

1 NONGLOSS LIGHT GREY-GREEN COATED REINFORCED CARBON-CARBON (RCC)

2 SEMIGLOSS BLACK-GREY COATED HIGH-TEMPERATURE REUSABLE SURFACE INSULATION (HRSI)

3 NONGLOSS OFF-WHITE (GREY-IVORY) COATED LOW-TEMPERATURE REUSABLE SURFACE INSULATION (LRSI)

4 SEMIGLOSS OFF-WHITE (GREY-IVORY) COATED FELT REUSABLE SURFACE INSULATION (FRSI)

5 GLOSSY LIGHT BLUE (SIMULATED WINDOWS)

6 SEMIGLOSS GREY METAL) MAIN ENGINES, ORBITAL MANEUVERING SUBSYSTEM (ENGINE NOZZLES ONLY), ELEVON HINGE DOORS, AND RUDDER HINGE SEAL

7 NONGLOSS BLACK REACTION CONTROL SUBSYSTEM ENGINE NOZZLE INTERIORS

8 MIRROR-REFLECTIVE FINISH (BRIGHT SILVER) ON INSIDE SURFACES OF PAYLOAD BAY DOORS

9 SEMIGLOSS WHITE TEFLON-COATED FABRIC PAYLOAD BAY LINER

EXTERNAL TANK AND SOLID-ROCKET BOOSTERS

11 NONGLOSS WHITE LO₂-LH₂ EXTERNAL TANK INSULATION

12 SEMIGLOSS WHITE PAINTED SOLID-ROCKET BOOSTERS (SRB'S)

13 NONGLOSS LIGHT-TAN SILICA PHENOLIC SRB ROCKET NOZZLES

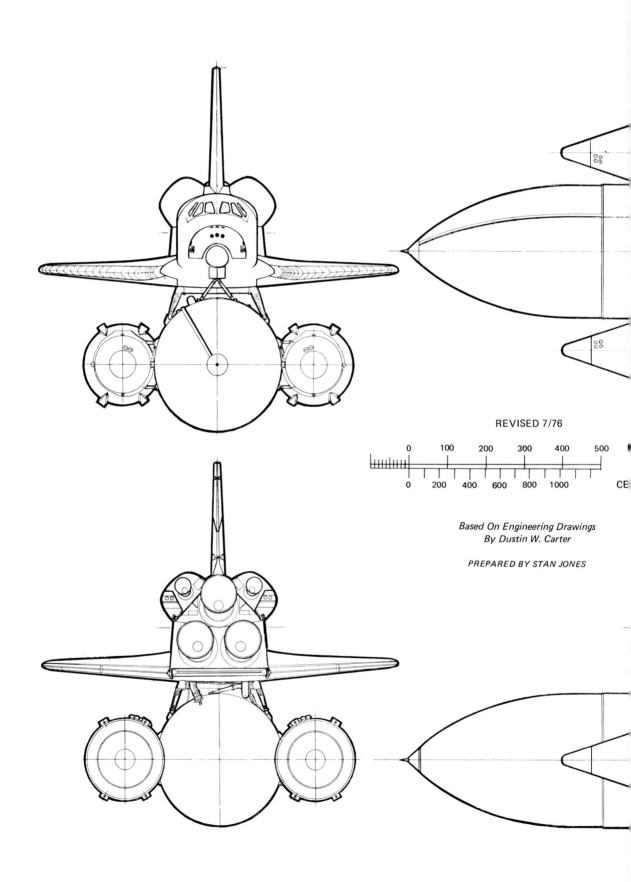

REVISED 7/76

Based On Engineering Drawings
By Dustin W. Carter

PREPARED BY STAN JONES

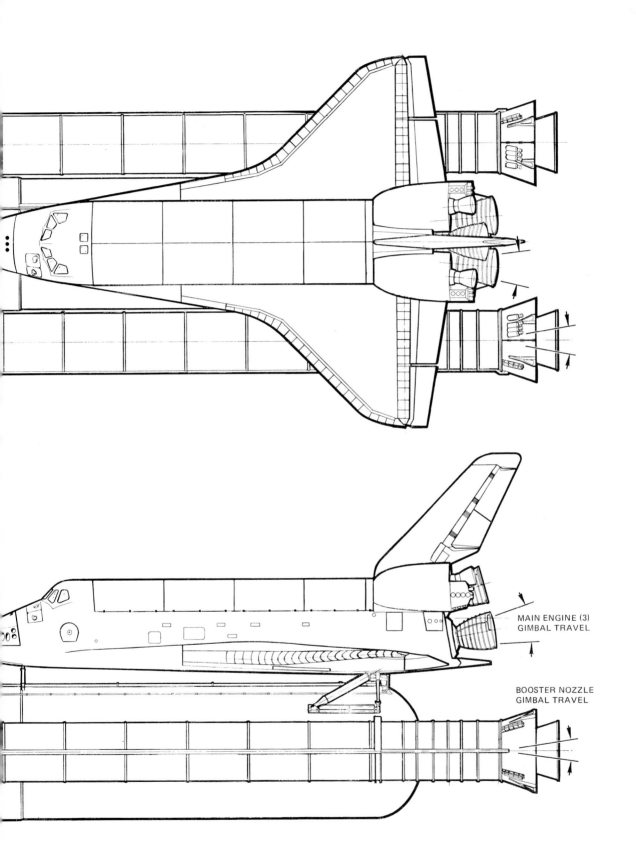

MAIN ENGINE (3)
GIMBAL TRAVEL

BOOSTER NOZZLE
GIMBAL TRAVEL

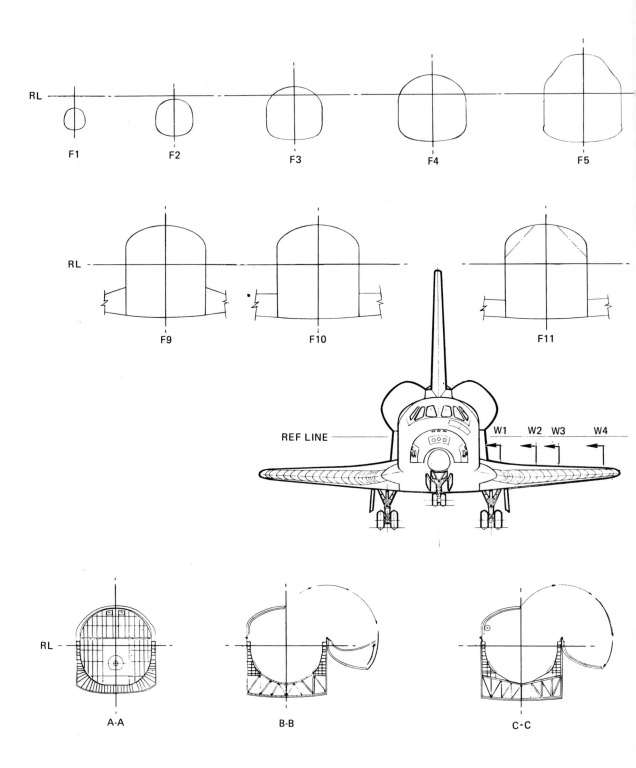

RL

F1   F2   F3   F4   F5

RL

F9   F10   F11

REF LINE   W1 W2 W3 W4

RL

A-A   B-B   C-C

SPACE SHUTTLE

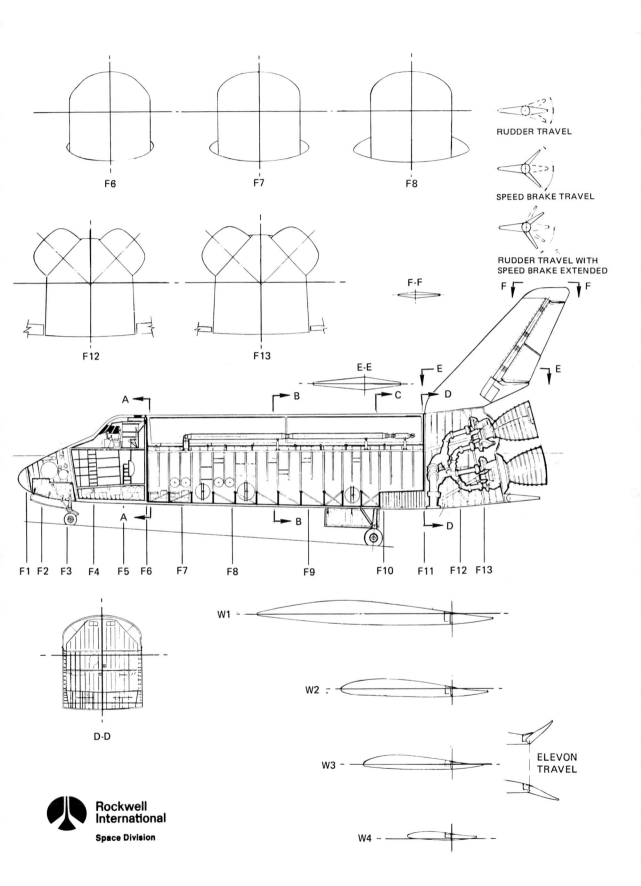

F6　　　　F7　　　　F8

RUDDER TRAVEL

SPEED BRAKE TRAVEL

RUDDER TRAVEL WITH
SPEED BRAKE EXTENDED

F12　　　　F13

F-F

E-E

A　　B　　C　　D　　E

F1 F2 F3 F4 F5 F6 F7 F8 F9 F10 F11 F12 F13

D-D

W1

W2

W3

ELEVON
TRAVEL

W4

**Rockwell
International**

Space Division

APPENDIX C

ASTRONAUT CANDIDATES
FOR THE SPACE SHUTTLE

By the end of 1977, thirty-five astronaut candidates were selected for the Space Shuttle program. On January 16, 1978, NASA Administrator, Dr. Robert A. Frosch, announced the selection of these men and women. A total of 8,079 applications were received during the year long recruiting period which ended on June 30, 1977. Between August and November of that year a total of 308 finalists were interviewed and underwent medical examinations at the Johnson Space Center in Houston. The newly selected candidates included 14 civilians and 21 military officers. For the first time women were selected as astronaut candidates; there were six in this group.

The group of candidates reported to Houston on July 1, 1978, to begin a two-year training program. Successful candidates joined the 27 astronauts already on active status in 1980. The candidates have been separated into two categories: Pilot and Mission Specialist. Pilots operate the Orbiter, maneuver it in Earth orbit and fly it to a runway landing. Mission Specialists have overall responsibility in cooperation with the Commander and pilot for mission success, management of consumable quantities, and a variety of activities related to experiment operations. This new group included 15 Pilot candidates and 20 Mission Specialist candidates.

A comprehensive training program was developed to qualify this group as astronauts by mid-1980. The program

consisted of classroom work, on-the-job-training assignments, and check out in T-38A jet trainer aircraft. The candidates became familiar with the overall NASA organization, its history and future plans and goals. They actively support NASA's public affairs program and are familiar with all aspects of the Space Transportation System. Pilot candidates are current in the T-38, while Mission Specialists are qualified as "rear seat" operators to handle flight planning, communications, and navigation. They received training in aerospace technology associated with the Space Shuttle and are ready to support technical or scientific assignments. Last, but not least, they are familiar with NASA's Civil Service regulations, astronaut standard operating procedures, and the code of conduct. To accomplish this training, over 300 hours of basic classroom work were scheduled during the first three months of the program. The T-38A check out began early in this two-year period. The astronauts use these aircraft to maintain proficiency in high performance aircraft and to travel between NASA and industry facilities as their duties require. Fifty-eight hours were scheduled for the check out of jet fighter pilot types with seventy-four hours planned for the nonfighter pilot with jet experience. Mission Specialists received more than seventy-four hours to qualify them for rear seat operations of a T-38A.

Most of the first three months of the program were spent in the classroom. The balance of the two-year period consisted of direct interaction with the Space Transportation System program. The candidates were given assignments by a supervising astronaut throughout their training and evaluation period. Such assignments helped to develop skills required for flight crew status as well as provide a basis for candidates to contribute productively to the orbital flight tests and development of Shuttle payloads.

An astronaut candidate assessment board, which is made up of the training program manager and individuals from the astronaut office, conducted quarterly progress reviews and passed on the final fitness of the 35 trainees. All went well, and in July 1980, NASA graduated 15 new astronaut Pilots and 20 new astronaut Mission Specialists. Beyond this point, there was still more training. Although the content and duration of such training was not fixed, it is necessary to operate the Space Transportation System elements common to all flights. Once a flight is assigned, the selected astronaut begins intensified training for that particular flight's mission. This may be an orbit insertion, on-orbit repair, or satellite retrieval mission. After each flight the astronaut is recycled into a pool of qualified flight personnel to await a new assignment. Between flights each astronaut is assigned other duties relevant to the Space Shuttle. In addition, they periodically receive recurring training to keep them proficient.

A short biographical description of each of the original 35 Shuttle candidates is presented below.

NAME: Guion S. Bluford, Jr., MAJ, USAF (PhD) — Mission Specialist

BIRTH: November 22, 1942, Philadelphia, PA

EDUCATION: Overbrook Senior High School, Philadelphia, PA; BS, Aerospace Engineering, Pennsylvania State University, 1964; MS, Aerospace Engineering, Air Force Institute of Technology, 1974; PhD, Aerospace Engineering, Air Force Institute of Technology, 1977

PRESELECTION POSITION: Chief, Aerodynamics and Airframe Branch, Aeromechanics Division, Air Force Dynamics Laboratory, Wright-Patterson AFB, OH

NAME: Daniel C. Brandenstein, LCDR, USN — Pilot

BIRTH: January 17, 1943, Watertown, WI

EDUCATION: Watertown High School, Watertown, WI; BS, Mathematics/Physics, University of Wisconsin, 1965

PRESELECTION POSITION: Naval Aviator and Maintenance Officer, Attack Squadron One Four Five, NAS Whidbey Island, Oak Harbor, WA

NAME: James F. Buchli, CAPT, USMC — Mission Specialist

BIRTH: June 20, 1945, New Rockford, SD

EDUCATION: Fargo Central High School, Fargo, ND; BS, USN Academy, 1967; MS, Aeronautical Systems, University of West Florida, 1975

PRESELECTION POSITION: Student, U.S. Naval Flight Test Engineering School, Patuxent River, MD

NAME: Michael L. Coats, LCDR, USN — Pilot

BIRTH: January 16, 1946, Sacramento, CA

EDUCATION: Ramona High School, Riverside, CA; BS, USN Academy, 1968; MS, Admin. of Science & Technology, George Washington University, 1977

PRESELECTION POSITION: Student, U.S. Navy Postgraduate School, Monterey, CA

NAME: Richard O. Covey, MAJ, USAF — Pilot

BIRTH: August 1, 1946, Fayetteville, AR

EDUCATION: Choctawhatchee High School, Shalimar, FL; BS, USAF Adademy, 1968; MS, Aeronautical/Astronautical Engineering, Purdue University, 1969

PRESELECTION POSITION: Commander, F-15 Joint Test Force, Air Force Test Center Detachment 2, Eglin AFB, FL

NAME: John O. Creighton, LCDR, USN — Pilot

BIRTH: April 28, 1943, Orange, TX

EDUCATION: Ballard High School, Seattle, WA; BS, USN Academy, 1966

PRESELECTION POSITION: Test Pilot, Naval Air Test Center, Patuxent River, MD

NAME: John M. Fabian, MAJ, USAF (PhD) — Mission Specialist

BIRTH: January 28, 1939, Goosecreek, TX

EDUCATION: Pullman High School, Pullman, WA; BS, Mechanical Engineering, Washington State University, 1962; MS, Aerospace Engineering, Air Force Institute of Technology, 1964; PhD, Aeronautics/Astronautics, University of Washington, 1974

PRESELECTION POSITION: Assistant Professor of Aeronautics, USAF Academy, CO

NAME: Anna L. Fisher, MD — Mission Specialist

BIRTH: August 24, 1949, Albany, NY

EDUCATION: San Pedro High School, San Pedro, CA; BS, Chemistry, University of California, Los Angeles, 1971; MD, University of California, Los Angeles, School of Medicine, 1976

PRESELECTION POSITION: Physician, Los Angeles, CA

NAME: Dale A. Gardner, LT, USN — Mission Specialist

BIRTH: November 8, 1948, Fairmont, MN

EDUCATION: Savanna Community High School, Savanna, IL; BS, Engineering Physics, University of Illinois, 1970

PRESELECTION POSITION: Naval Flight Officer, Air Test and Evaluation Squadron Four, NAS Point Mugu, CA

NAME: Robert L. Gibson, LT, USN — Pilot

BIRTH: October 30, 1946, Cooperstown, NY

EDUCATION: Huntington High School, Huntington, NY; BS, Aeronautical Engineering, California Polytechnic State Univ., 1969

PRESELECTION POSITION: Test Pilot, Naval Air Test Center, Patuxent River, MD

NAME: Frederick D. Gregory, MAJ, USAF — Pilot

BIRTH: January 7, 1941, Washington, DC

EDUCATION: Anacosta High School, Washington, DC; BS, USAF Academy, 1964; MS, Information Systems, George Washington University, 1977

PRESELECTION POSITION: Armed Forces Staff College, Norfolk, VA

NAME: Stanley D. Griggs — Pilot

BIRTH: September 7, 1939, Portland, OR

EDUCATION: Lincoln High School, Portland, OR; BS, USN Academy, 1962; MSA, Management Engineering, George Washington University, 1970

PRESELECTION POSITION: Chief, Shuttle Training Aircraft Operations Office, NASA/Johnson Space Center, Houston, TX

NAME: Terry J. Hart — Mission Specialist

BIRTH: October 27, 1946, Pittsburgh, PA

EDUCATION: Mt. Lebanon High School, Pittsburgh, PA; BS, Mechanical Engineering, Lehigh University, 1968; MS, Mechanical Engineering, Massachusetts Institute of Technology, 1969

PRESELECTION POSITION: Technical Staff Member, Bell Telephone Laboratories, Whippany, NJ

NAME: Frederick H. Hauck, CDR, USN — Pilot

BIRTH: April 11, 1941, Long Beach, CA

EDUCATION: St. Albans High School, Mt. St. Alban, Washington, DC; BS, General Physics, Tufts University, 1962; MS, Nuclear Engineering, Massachusetts Institute of Technology, 1966

PRESELECTION POSITION: Executive Officer, Attack Squadron One Four Five, NAS Whidbey Island, Oak Harbor, WA

NAME: Steven A. Hawley, PhD — Mission Specialist

BIRTH: December 12, 1951, Ottawa, KS

EDUCATION: Salina Central High School, Salina, KS; BA, Astronomy and Physics, University of Kansas, 1973; PhD, Astronomy, University of California, Santa Cruz, 1977

PRESELECTION POSITION: Postdoctoral Research Associate, Cerro Tololo Inter-American Observatory, La Serena, Chile

---

NAME: Jeffrey A. Hoffman, PhD — Mission Specialist

BIRTH: November 2, 1944, New York, NY

EDUCATION: Scarsdale High School, Scarsdale, NY; BA, Astronomy, Amherst College, 1966; PhD, Astrophysics, Harvard University, 1971

PRESELECTION POSITION: Astrophysics Research Staff, Massachusetts Institute of Technology, Center for Space Research, Cambridge, MA

NAME: Shannon W. Lucid, PhD — Mission Specialist

BIRTH: January 14, 1943, Shanghai, China

EDUCATION: Bethany High School, Bethany, OK; BS, Chemistry, University of Oklahoma, 1963; MS, Biochemistry, University of Oklahoma, 1970; PhD, Biochemistry, University of Oklahoma, 1973

PRESELECTION POSITION: Postdoctoral Fellow, Oklahoma Medical Research Foundation, Oklahoma City, OK

NAME: Jon A. McBride, LCDR, USN — Pilot

BIRTH: August 14, 1943, Charleston, WV

EDUCATION: Woodrow Wilson High School, Beckley, WV; BS, Aeronautical Engineering, USN Postgraduate School, 1971

PRESELECTION POSITION: Test Pilot, Air Test and Evaluation Squadron Four, Point Mugu, CA

NAME: Ronald E. McNair, PhD — Mission Specialist

BIRTH: October 21, 1950, Lake City, SC

EDUCATION: Carver High School, Lake City, SC; BS, Physics, North Carolina A & T University, 1971; PhD, Physics, Massachusetts Institute of Technology, 1977

PRESELECTION POSITION: Member of the

Technical Staff, Optical Physics Department, Hughes Research Laboratories, Malibu, CA

NAME: Richard M. Mullane, CAPT, USAF — Mission Specialist

BIRTH: Sepember 10, 1945, Wichita Falls, TX

EDUCATION: St. Pius X High School, Albuquerque, NM; BS, U. S. Military Academy, 1967; MS, Aeronautical Engineering, Air Force Institute of Technology, 1975

PRESELECTION POSITION: Flight Test Weapon Systems Operator, 3246th Test Wing, Eglin AFB, FL

NAME: Steven R. Nagel, CAPT. USAF — Pilot

BIRTH: October 27, 1946, Canton, IL

EDUCATION: Canton High School, Canton, IL; BS, Aeronautical/Astronautical Engineering, University of Illinois, 1969

PRESELECTION POSITION: Test Pilot, Air Force Flight Test Center, Edwards AFB, CA

NAME: George D. Nelson, PhD — Mission Specialist

BIRTH: July 13, 1950, Charles City, IA

EDUCATION: Willmar Senior High School, Willmar, MN; BS, Physics, Harvey Mudd University, 1972; MS, Astronomy, University of Washington, 1974; PhD, Astronomy, University of Washington, 1977

PRESELECTION POSITION: Research Associate, Astronomy Department, University of Washington, Seattle, WA

NAME: Ellison S. Onizuka, CAPT, USAF — Mission Specialist

BIRTH: June 24, 1946, Kealakekua, HI

EDUCATION: Konawaena High School, Kealakekua, HI; BS, Aerospace Engineering, University of Colorado, 1969; MS, Aerospace Engineering, University of Colorado, 1969

PRESELECTION POSITION: Chief, Engineering Support Section, Training Resources Branch, USAF Test Pilot School, Edwards AFB, CA

NAME: Judith A. Resnik, PhD — Mission Specialist

BIRTH: April 5, 1949, Akron, OH

EDUCATION: Firestone High School, Akron, OH; BS, Electrical Engineering, Carnegie-Mellon University, 1970; PhD, Electrical Engineering, University of Maryland, 1977

PRESELECTION POSITION: Engineering Staff, Product Development, Xerox Corporation, El Segundo, CA

NAME: Sally K. Ride — Mission Specialist

BIRTH: May 26, 1951, Los Angeles, CA

EDUCATION: Westlake High School, Los Angeles, CA; BS, Physics, Stanford University, 1973; BA, English, Stanford University, 1973; MS, Physics, Stanford University, 1975

PRESELECTION POSITION: Research Assistant, Physics Department, Stanford University, Stanford, CA

NAME: Francis R. Scobee, MAJ, USAF — Pilot

BIRTH: May 19, 1939, Cle Elum, WA

EDUCATION: Auburn High School, Auburn, WA; BS, Aerospace Engineering, University of Arizona, 1965

PRESELECTION POSITION: Test Pilot, Air Force Flight Test Center, Edwards AFB, CA

NAME: Margaret R. Seddon, MD — Mission Specialist

BIRTH: November 8, 1947, Murfreesboro, TN

EDUCATION: Central High School, Murfreesboro, TN; BA, Physiology, University of California, Berkeley, 1970; MD, University of Tennessee College of Medicine, 1973

PRESELECTION POSITION: Resident Physician, Department of Surgery, City of Memphis Hospital, Memphis, TN

NAME: Brewster H. Shaw, Jr., CAPT, USAF — Pilot

BIRTH: May 16, 1945, Cass City, MI

EDUCATION: Cass City High School, Cass City, MI; BS, Engineering Mechanics, University of Wisconsin, 1968; MS, Engineering Mechanics, University of Wisconsin 1969

PRESELECTION POSITION: Instructor, U. S. Air Force Test Pilot School, Edwards AFB, CA

NAME: Loren J. Shriver, CAPT, USAF — Pilot

BIRTH: September 23, 1944, Jefferson, IA

EDUCATION: Paton Consolidated High School, Paton, IA; BS, USAF Academy, 1967; MS, Astronautics, Purdue University, 1968

PRESELECTION POSITION: Test Pilot, Air Force Flight Test Center, Edwards AFB, CA

NAME: Robert L. Stewart, MAJ, U. S. ARMY — Mission Specialist

BIRTH: August 13, 1942, Washington, DC

EDUCATION: Hattiesburg High School, Hattiesburg, MS; BS, Mathematics, University of Southern Mississippi, 1964; MS, Aerospace Engineering, University of Texas, Arlington, 1971

PRESELECTION POSITION: Test Pilot, U. S. Army Aviation Engineering Flight Activity, Edwards AFB, CA

NAME: Kathryn D. Sullivan, Ph.D. — Mission Specialist

BIRTH: October 3, 1951, Paterson, NJ

EDUCATION: Taft High School, Woodland Hills, CA; BS, Earth Sciences, University of California, Santa Cruz, 1973; PhD, Geology, Dalhousie University, Halifax, Nova Scotia, 1978

PRESELECTION POSITION: Postgraduate Student, National Research Council, Dalhousie University, Halifax, Nova Scotia, Canada

---

NAME: Norman E. Thagard, MD — Mission Specialist

BIRTH: July 3, 1943, Marianna, FL

EDUCATION: Paxon High School, Jacksonville, FL; BS, Engineering Science, Florida State University, 1965; MS, Engineering Science, Florida State University, 1966; MD, University of Texas Southwestern Medical School, 1977

PRESELECTION POSITION: Intern, Department of Internal Medicine, Medical University of South Carolina, Charleston, SC

NAME: James D. van Hoften, PhD — Mission Specialist

BIRTH: June 11, 1944, Fresno, CA

EDUCATION: Mills High School, Millbrae, CA;

BS, Civil Engineering, University of California, Berkeley, 1966; MS, Hydraulic Engineering, Colorado State University, 1968; PhD, Fluid Mechanics, Colorado State University, 1976

PRESELECTION POSITION: Assistant Professor of Civil Engineering, University of Houston, Houston, TX

NAME: David M. Walker, LCDR, USN — Pilot

BIRTH: May 20, 1944, Columbus, GA

EDUCATION: Eustis High School, Eustis, FL; BS, USN Academy, 1966

IN; BS, Mechanical Engineering, Purdue University, 1964

PRESELECTION POSITION: Naval Aviator, USS America

NAME: Donald E. Williams, LCDR, USN — Pilot

BIRTH: February 13, 1942, Lafayette, IN

EDUCATION: Otterbein High School, Otterbein, IN

PRESELECTION POSITION: Naval Aviator, Readiness Training Squadron, NAS Lemoore, CA

Thirty-five astronaut candidates have finished a two-year training program to become Space Shuttle crewmembers. The group includes 15 pilot candidates and 20 Mission Specialist candidates.

BLUFORD	BRANDENSTEIN	BUCHLI	COATS	COVEY	CREIGHTON	FABIAN
FISHER	GARDNER	GIBSON	GREGORY	GRIGGS	HART	HAUCK
HAWLEY	HOFFMAN	LUCID	McBRIDE	McNAIR	MULLANE	NAGEL
NELSON	ONIZUKA	RESNIK	RIDE	SCOBEE	SEDDON	SHAW
SHRIVER	STEWART	SULLIVAN	THAGARD	VAN HOFTEN	WALKER	WILLIAMS

In May 1980, an additional 19 astronaut candidates were selected for the Shuttle program. Their training began on July 1, 1980 and lasted 12 months A brief set of comments appears below on each.

Sixteen of the 19 astronaut candidates selected in May 1980 are shown here with two European trainees as payload specialists at the Johnson Space Center. Kneeling, left to right, on the front row, are Claude Nicollier and Ubbo Ochels, payload specialists; Richard N. Richards, Sherwood C. Spring, Roy D. Bridges, David C. Hilmers and Charles F. Bolden. Standing left to right, are Robert C. Springer, Michael J. Smith, John M. Lounge, Bonnie J. Dunbar, Jerry L. Ross, Mary L. Cleave, Franklin R. Chang, John E. Blaha, William F. Fisher, James P. Bagian and Bryan D. O'Conner. Not appearing are Guy S. Gardner, Ronald J. Grabe and David C. Leestma.

NAME: James P. Bagian (MD)—Mission
Specialist
BIRTH: February 22, 1952, Philadelphia, PA
EDUCATION: Central High School, Philadelphia, PA; BS, Mechanical Engineering, Drexel University, 1973; MD, Thomas University, 1977

NAME: John E. Blaha (Colonel, USAF)—Pilot
BIRTH: August 26, 1942, San Antonio, TX
EDUCATION: Granby High School, Norfolk, VA; BS, Engineering Science, USAF Academy, 1965; MS, Astronautical Engineering, Purdue University, 1966

NAME: Charles F. Bolden, Jr. (Major, USMC)—
Pilot

BIRTH: August 19, 1946, Columbia, SC
EDUCATION: C.A. Johnson High School, Columbia, SC; BS, Electrical Science, US Naval Academy, 1968; MS, Systems Management, University of Southern California, 1978

NAME: Roy D. Bridges, Jr. (Lieutenant Colonel, USAF)—Pilot
BIRTH: July 19, 1943, Atlanta, GA
EDUCATION: Gainesville High School, Gainesville, GA; BS, Engineering Science, USAF Academy, 1965; MS, Astronautics, Purdue University 1966

NAME: Franklin R. Chang (PhD)—Mission
Specialist
BIRTH: April 5, 1950, San Jose, Costa Rica

EDUCATION: Hartford High School, Hartford, CT; BS, Mechanical Engineering, University of Connecticut, 1973; PhD, Applied Plasma Physics, Massachusetts Institute of Technology, 1977

NAME: Mary L. Cleave (PhD)—Mission Specialist
BIRTH: February 5, 1947, Southampton, NY
EDUCATION: **Great Neck North High School,** Great Neck, NY; BS, Biological Sciences, Colorado State University, 1969; MS, Microbial Ecology, Utah State University, 1975; PhD, Civil and Environmental Engineering, Utah State University, 1979

NAME: Bonnie J. Dunbar (MS)—Mission Specialist
BIRTH: March 3, 1949, Sunnyside, WA
EDUCATION: Sunnyside High School, Sunnyside, WA; BS, MS, Ceramic Engineering, University of Washington, 1971, 1975; doctoral candidate, Biomedical Engineering, University of Houston

---

NAME: William F. Fisher (MD)—Mission Specialist
BIRTH: April 1, 1946, Dallas, TX
EDUCATION: North Syracuse Central High School, North Syracuse, NY; BA, Pre-Med, Stanford University, 1968; MD, University of Florida, 1975; MS, Engineering, University of Houston, 1980

NAME: Guy S. Gardner (Major, USAF)—Pilot
BIRTH: January 6, 1948, Alto Vista, VA
EDUCATION: George Washington High School, Alexandria, VA; BS, Engineering Sciences (Astronautics and Mathematics), USAF Academy, 1969; MS, Aeronautics and Astronautics, Purdue University, 1970

NAME: Ronald J. Grabe (Major, USAF)—Pilot
BIRTH: June 13, 1945, New York, NY
EDUCATION: Stuyvesant High School, New York, NY; BS, Engineering Science, USAF Academy, 1966; Fulbright Scholar, Aeronautics, Technische Hochschule, Darmstadt, West Germany, 1967

NAME: David C. Hilmers (Captain, USMC)—Mission Specialist
BIRTH: January 28, 1950, Clinton, IA
EDUCATION: Central Community High School, DeWitt, IA; BS, Mathematics, Cornell College, 1972; MS, Electrical Engineering, US Naval Postgraduate School, 1978

NAME: David C. Leestma (Lieutenant Commander, USN)—Mission Specialist
BIRTH: May 6, 1949, Muskegon, MI
EDUCATION: Tustin High School, Tustin, CA; BS, Aeronautical Engineering, US Naval Academy, 1971; MS, Aeronautical Engineering, US Naval Postgraduate School, 1972

NAME: John M. Lounge (MS)—Mission Specialist
BIRTH: June 28, 1946, Denver, Colorado
EDUCATION: Burlington High School, Burlington, CO; BS, Physics and Mathematics, US Naval Academy, 1969; MS, Astrogeophysics, University of Colorado, 1970; advanced degree candidate, Operations Research (Human Factors Engineering), University of Houston

---

NAME: Bryan D. O'Connor, (Major, USMC)—Pilot
BIRTH: September 6, 1946, Orange, CA
EDUCATION: Twenty Nine Palms High School, Twenty Nine Palms, CA; BS, Engineering, US Naval Academy, 1968; MS, Aeronautical Systems, University of West Florida, 1970

NAME: Richard N. Richards (Lieutenant Commander, USN)—Pilot
BIRTH: August 24, 1946, Key West, FL
EDUCATION: Riverview Gardens High School, St. Louis, MS; BS, Chemical Engineering, University of Missouri, 1969; MS, Aeronautical Systems, University of West Florida, 1970

NAME: Jerry L. Ross (Captain, USAF)—Mission Specialist
BIRTH: January 20, 1948, Crown Point, IN
EDUCATION: Crown Point High School, Crown Point, IN; BS, MS, Mechanical Engineering,

Purdue University, 1970, 1972

NAME: Michael J. Smith (Lieutenant Commander, USN)—Pilot
BIRTH: April 30, 1945
EDUCATION: Beaufort High Schol, Beaufort, NC; BS, Naval Science, US Naval Academy, 1967; MS, Aeronautical Engineering, US Naval Postgraduate School, 1968

NAME: Robert C. Springer (Lieutenant Colonel, USMC)—Mission Specialist

BIRTH: May 21, 1942, St. Louis, MS
EDUCATION: Ashland High School, Ashland, OH; BS, Naval Science, US Naval Academy 1964; MS, Operations Research and Systems Analysis, US Naval Postgraduate School, 1971

NAME: Sherwood C. Spring (Major, USA)—Mission Specialist
BIRTH: September 3, 1944, Hartford, CT
EDUCATION: Ponagansett High School, Chepacket, RI; BS, General Engineering, US Military Academy, 1967; MS, Aerospace Engineering, University of Arizona, 1974

APPENDIX D

GLOSSARY OF THE SPACE TRANSPORTATION SYSTEM

aft flight deck — That part of the Orbiter cabin on the upper deck where payload controls are located.

airlock — A compartment, capable of being depressurized without depressurization of the Orbiter cabin, used to transfer crewmembers and equipment.

barbecue mode — Orbiter in slow rotation for thermal conditioning.

capture — The maneuver of the Remote Manipulator System when making contact with and firmly attaching to a free-flying object.

cargo — The total complement of payloads (one or more) on any one flight. It includes everything contained in the Orbiter cargo bay plus other equipment, hardware, and consumables located elsewhere in the Orbiter that are user-unique.

cargo bay — The unpressurized mid-part of the Orbiter fuselage behind the crew cabin where payloads are carried. Its maximum usable envelope is 15 feet in diameter and 60 feet long. Hinged doors extend the full length of the bay.

cargo bay liner — Protective soft material used to isolate sensitive payloads from the bay structure.

Commander — This crewmember has ultimate responsibility for the safety of embarked personnel and has authority throughout the flight to deviate from the flight plan, procedures, and crew assignments as necessary to preserve human safety or vehicle integrity. The commander is also responsible for the overall execution of the flight plan.

deployment — The process of removing a payload from a stowed or berthed position in the cargo bay and releasing that payload to a position free of the Orbiter.

European Space Agency — An international organization acting on behalf of its member states (Belgium, Denmark, France, Federal Republic of Germany, Italy, the Netherlands, Spain, Sweden, Switzerland, and the United Kingdom). The ESA directs a European industrial team responsible for the development and manufacture of Spacelab.

experimenter — A user of the Space Transportation System who ordinarily will be an individual whose experiment is a small part of the total payload.

External Tank — Element of the Space Shuttle system that contains liquid propellants for the Orbiter main engines. It is jettisoned prior to orbit insertion.

extravehicular activity — Activities by crewmembers conducted outside the pressurized portion of the Orbiter.

Extravehicular Mobility Unit — A self-contained (no umbilicals) life support system and anthropomorphic pressure garment for use by crewmembers during extravehicular activities. It also provides thermal and micrometeoroid protection.

flight control team — A group of ground controllers at the Mission Control Center on duty to provide real-time support for the duration of each Shuttle flight.

flight data file — The onboard complement of crew activity plans, procedures, reference material, and test data available to the crew for flight execution. There will normally be one flight data file for crew activities and one for payload activities.

flight kit—Optional hardware (including consumables) to provide additional, special, or extended services to payloads. Kits are packaged in such a way that they can be installed and removed easily.

flight phases — Prelaunch, launch, in orbit, deorbit, entry, landing, and postlanding.

free flying system — Any satellite or payload that is detached from the Orbiter during operational phases and is capable of independent operation.

inclination — The angle between the orbit plane and the equatorial plane. It corresponds to the highest latitude over which a satellite passes.

instrument pointing subsystem — Spacelab hardware and software for precision pointing and stability of experiment equipment.

intact abort — Any of three abort modes which are designed to bring the Orbiter and crew back to a safe landing.

Inertial Upper Stage — Solid propulsive upper stage designed to place spacecraft in high Earth and retrieved by the Remote Manipulator System.

launch azimuth — Direction of ground track after leaving the launch pad, measured clockwise from true North.

launch pad — The area at which the stacked Space Shuttle undergoes final prelaunch checkout and countdown and from which it is launched.

Long Duration Exposure Facility — Free-flying reusable satellite designed primarily for small passive or self-contained active experiments that require prolonged exposure to space. It is launched in the Orbiter cargo bay and deployed and retrieved by the Remote Manipulator System.

Manned Maneuvering Unit — A propulsive backpack device for maneuvering during extravehicular activities. It uses a low-thrust, dry, cold nitrogen propellant.

Mission Control Center — Central area at the Johnson Space Center for control and support of all phases of Shuttle flights.

Mission Specialist — This crewmember is responsible for coordination of all payload operations and directs the allocation of resources to accomplish mission objectives. The mission specialist will have prime responsibility for experiments to which no Payload Specialist is assigned, and will assist a Payload Specialist when appropriate.

mission station — Location on the Orbiter aft flight deck from which payload operations are performed, usually by the Mission Specialist.

mixed payloads — Cargo containing more than

one type of payload.

Mobile Launch Platform — The structure on which the elements of the Space Shuttle are stacked in the Vehicle Assembly Building and are moved to the launch pad.

mobility aid — Handrails or footrails to help crewmembers move about the spacecraft.

Multimission Modular Spacecraft — Free-flying system built in sections so that it can be adapted to many missions requiring Earth-orbiting remote-sensing spacecraft. It is launched in the Orbiter cargo bay and deployed and retrieved by the Remote Manipulator System.

orbital flight test — One of several scheduled developmental space flights of the Space Shuttle System.

Orbital Maneuvering Subsystem — Orbiter engines that provide the thrust to perform orbit insertion, circularization, or transfer; rendezvous; and deorbit.

Orbiter — Manned orbital flight vehicle of the Space Shuttle System.

Orbiter Processing Facility — Building near the Vehicle Assembly Building at the Kennedy Space Center with two bays in which Orbiters undergo postflight inspection, maintenance, and premate checkout prior to payload installation. Payloads are also installed horizontally into the Orbiter in this building.

pallet — An unpressurized platform, designed for installation in the Orbiter cargo bay, for mounting instruments and equipment requiring direct space exposure.

payload canister — Environmentally controlled transporter for use at the launch site. It is the same size and configuration as the Orbiter cargo bay.

Payload Changeout Room — An environmentally controlled room at the launch pad for inserting payloads vertically into the Orbiter cargo bay.

Payload Specialist — This crewmember, who may or may not be a career astronaut, is responsible for the operation and management of the experiments or other payload elements that are assigned to him or her, and for the achievement of their objectives. The payload specialist will be an expert in experiment design and operation.

Pilot — This crewmember is second in command of the flight and assists the Commander as required in the conduct of all phases of Orbiter flight.

Reaction Control Subsystem — Thrusters on the

Orbiter that provide attitude control and three-axis translation during orbit insertion, on-orbit, and reentry phases of flight.

Remote Manipulator System — Mechanical arm on the cargo bay longeron. It is controlled from the Orbiter aft flight deck to deploy, retrieve, or move payloads.

retrieval — The process of utilizing the Remote Manipulator System and/or other handling aids to return a captured payload to a stowed or berthed position. No payload is considered retrieved until it is fully stowed for safe return or berthed for repair and maintenance tasks.

simulator — A heavily computer-dependent training facility that imitates flight hardware responses.

Solid Rocket Booster — Element of the Space Shuttle that consists of two solid rocket motors to augment ascent thrust at launch. They are separated from the Orbiter soon after lift-off and recovered for reuse.

Spacelab — A general-purpose orbiting laboratory for manned and automated activities in near-Earth orbit. It includes both module and pallet sections, which can be used separately or in several combinations.

Space Shuttle Vehicle — Orbiter, External Tank, and two Solid Rocket Boosters.

Space Tracking and Data Network — A number of ground-base stations having direct communications with NASA flight vehicles.

spinning solid upper stage — Propulsive upper stage designed to deliver spacecraft of the Delta and Atlas-Centaur classes to Earth orbits beyond the capabilities of the Space Shuttle.

tilt/spin table — Mechanism installed in Orbiter cargo bay that deploys the spinning solid upper stage with its spacecraft.

Tracking and Data Relay Satellite system — Two-satellite communications system providing principal coverage for geosynchronous orbit for all Shuttle flights.

user — An organization or individual requiring the services of the Space Transportation System.

Vehicle Assembly Building — High-bay building near the Kennedy Space Center launch complex in which the Shuttle elements are stacked onto the mobile launch platform. It is also used for vertical storage of the External Tanks.

Western Launch Operations Division — NASA's operation at Vandenberg Air Force Base.

APPENDIX E

ACRONYMS
TO LISTEN FOR

ALT	Approach and Landing Tests
BSM	Booster Separation Motor
DFRC	Dryden Flight Research Center
DOD	Department of Defense
EAFB	Edwards Air Force Base
EMU	Extravehicular Mobility Unit
EPA	Environmental Protection Agency
ESA	European Space Agency
ET	External Tank
EVA	Extravehicular Activity
FFTO	Free-flying Teleoperator
IDB	Insuit Drink Bag
IUS	Inertial Upper Stage
IVA	Intravehicular Activity
JSC	Johnson Space Center
KSC	Kennedy Space Center
MCC	Mission Control Center
MECO	Main Engine Cutoff
MMS	Multimission Modular Spacecraft
MMU	Manned Maneuvering Unit
MS	Mission Specialist
MSFC	Marshall Space Flight Center
NASA	National Aeronautics and Space Administration
OMS	Orbital Maneuvering Subsystem
OPF	Orbiter Processing Facility
OV	Orbiter Vehicle
PCR	Payload Changeout Room
PLSS	Primary Life Support System
POS	Portable Oxygen Supply
PRS	Personnel Rescue System
PS	Payload Specialist
RCS	Reaction Control System
RMS	Remote Manipulator System
RSI	Reusable Surface Insulation
RV	Rescue Vehicle
SRB	Solid Rocket Booster
SSME	Space Shuttle Main Engine
SSUS	Spinning Solid Upper Stage

STDN	Space Tracking and Data Network
STS	Space Transportation System
TAEM	Terminal Area Energy Management
TDRSS	Tracking and Data Relay Satellite System
TPS	Thermal Protection System
USAF	U. S. Air Force
VAB	Vehicle Assembly Building
VAFB	Vandenberg Air Force Base
WSMR	White Sands Missile Range

INDEX

United Technology Corporation, 10
upper stage, 96,110
UTC tugs, Freedom, 93
   Liberty, 93

Vandenberg Air Force Base, 22, 25, 28, 29, 31, 33,
   36, 38, 48
Vehicle Assembly Building, 30, 38, 40
via satellite, 156
von Braun, Wernher, 9, 139, 178

waste collection, 61, 126
water spray concept, 90

water suppression technique, 84
weather satellites, 35
weightlessness, 95
Westar, 161
Western Union, 161
White, Edward, 66
White Sands Missile Range, 25
Wright brothers, 7

Young, John W., 25, 29, 68

zero defects, 140

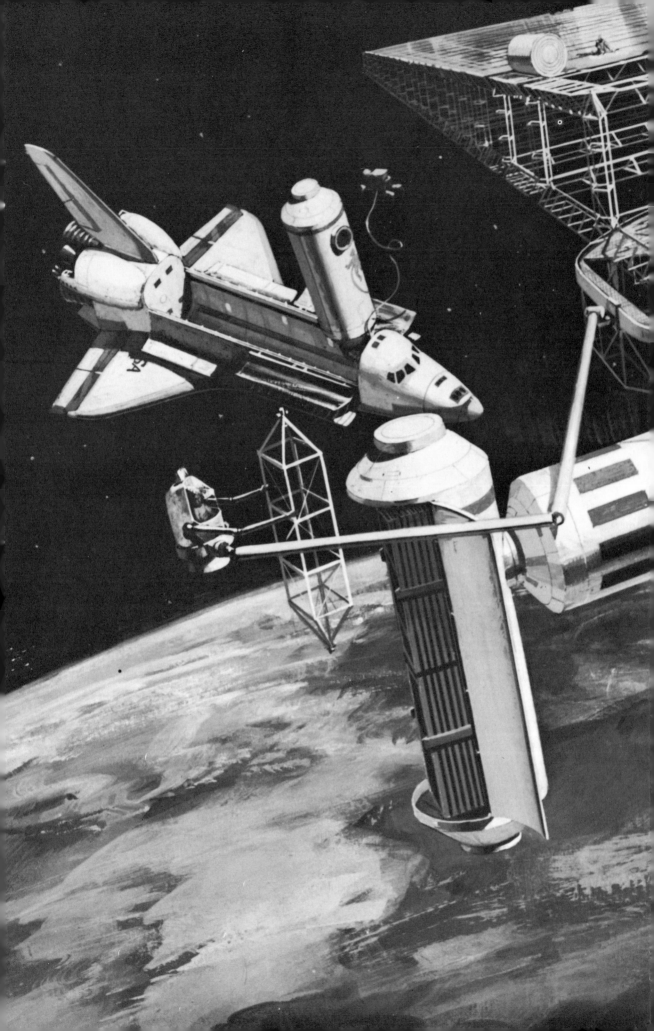